ENERGY
FOR
FUTURE
PRESIDENTS

ENERGY
FOR
FUTURE
PRESIDENTS

THE SCIENCE
BEHIND THE HEADLINES

RICHARD A.
MULLER

W. W. Norton & Company

New York • London

Copyright © 2012 by Richard A. Muller

All rights reserved
Printed in the United States of America
First Edition

For information about permission to reproduce selections from this book,
write to Permissions, W. W. Norton & Company, Inc.,
500 Fifth Avenue, New York, NY 10110

For information about special discounts for bulk purchases, please contact
W. W. Norton Special Sales at specialsales@wwnorton.com or 800-233-4830

Manufacturing by R R Donnelley, Harrisonburg, VA
Production manager: Anna Oler

Library of Congress Cataloging-in-Publication Data

Muller, R. (Richard)
Energy for future presidents : the science behind the headlines /
Richard A. Muller. — 1st ed.
p. cm.
Includes bibliographical references and index.
ISBN 978-0-393-08161-9 (hardcover)
1. Power resources—Social aspects. 2. Energy policy—Social aspects.
3. Technology and state. I. Title.
TJ163.2.M854 2012
333.79—dc23
                    2012015586

W. W. Norton & Company, Inc.
500 Fifth Avenue, New York, N.Y. 10110
www.wwnorton.com

W. W. Norton & Company Ltd.
Castle House, 75/76 Wells Street, London W1T 3QT

1 2 3 4 5 6 7 8 9 0

to ROSEMARY
for encouraging me
to hike off-trail

CONTENTS

ACKNOWLEDGMENTS

I AM GRATEFUL to several people for their helpful suggestions on an early draft of the manuscript, including Marlan Downey, Jonathan Katz, Jonathan Levine, and Dick Garwin. My foray into energy was inspired, in part, by similar forays by several friends, particularly Steve Koonin, Art Rosenfeld, Steve Chu, and Nate Lewis. It has been a great joy to work on energy technology and strategy with my daughter Elizabeth Muller, the cofounder of our Berkeley Earth Surface Temperature Project, and the CEO of our consulting company, Muller and Associates, on energy technology and strategy. I thank Stephanie Hiebert, my copy editor, for her patience and care. I am particularly grateful to Jack Repcheck both for inspiring me to write this book and for his invaluable guidance in both content and style.

PREFACE

The trouble with most folks isn't so much their ignorance;
it's know'n so many things that ain't so.

— *Josh Billings, a nineteenth-century humorist
but often misattributed to Mark Twain*[1]
(thus making it a self-illustrating aphorism)

A FUTURE president must understand energy. You know that. And, as the introductory aphorism implies, it is equally important that you be aware of the misunderstandings of everyone else. When you are president, you will have the primary responsibility for gently correcting the mistakes of the people who need to be convinced. You will have to be the public's energy instructor.

There is a lot to understand. When you put politics temporarily aside and look at energy in an objective way, you reach conclusions that are often counterintuitive and unexpected. Following is a list of some of the results that I'll discuss in this book. Even if you don't find them surprising yourself, most of your constituents will.

- The disasters of Fukushima and the Gulf oil spill were not nearly as catastrophic as many people think, and they should not imply any major change in energy policy.
- Global warming, although real and caused largely by humans, can be controlled only if we find inexpensive or profitable methods to reduce greenhouse gas emissions in China and the developing world.
- We have recently learned that we can exploit immense natural-gas reserves found in shale. It's not an exaggeration to call this discovery

a windfall. Shale gas will play a central role in US energy policy over the next few decades.

- The United States is running low not on fossil fuel, but only on transportation fuel. The keys to the future lie in synfuel (manufactured gasoline), natural gas, shale oil reserves, and improved automobile mileage.

- Energy productivity can be improved enormously. Investments in efficiency and conservation can yield returns much better than those of Bernard Madoff's Ponzi scheme, and what's more, the returns are tax-free.

- Solar energy is undergoing spectacular development, but its potential lies in solar cells, not in solar-thermal power plants. The main competitor to solar is natural gas.

- Wind has significant potential as a supplemental source of electricity, but it requires a better power transmission grid. Now that wind power production is growing rapidly, there is growing opposition from environmentalists. The main competitor to wind is natural gas.

- Energy storage (to address the intermittency of wind and solar) is a difficult and expensive problem. The most cost-effective approach is probably with batteries, although natural-gas backup might be cheaper.

- Nuclear power is safe, and waste storage is not a difficult problem. Fears are driven by unfamiliarity and misinformation. The main competitor to nuclear power is natural gas.

- The primary future value of biofuel will be for transportation energy security, not to prevent global warming. Corn ethanol should not be considered a biofuel. The main competitor to biofuel is natural gas.

- Synfuels are practical and important and, when developed, should keep the cost of oil to $60 per barrel or less. For automobile fuel, synfuel manufactured from natural gas is one of the few energy sources that might beat out compressed natural gas.

- There is a fast-breaking and potentially disruptive (in the good sense) new energy source: shale oil. The US reserves are enormous, and practical means of extraction have been developed. Shale oil appears to be cheaper to produce than synfuel, and it is likely to offer stiff competition to that industry.

- The hydrogen economy is going nowhere. Some of the most romantic alternative-energy sources, including geothermal, tidal energy, and wave energy, also have no large-scale future.

- Hybrid autos have a great future, but plug-in hybrids and all-electric automobiles do not; they cost much more to operate than do gasoline cars once you include battery replacement cost. There is an exception: autos that run on lead-acid batteries, with very short ranges (on the order of 40–60 miles), could achieve widespread use in China, India, and the rest of the developing world.

- Virtually none of the publicly proposed solutions to the danger of increased carbon dioxide, if implemented, have any realistic chance of working. "Setting an example" fails if the example is one that the developing world can't afford to follow. Perhaps the only workable solution is to encourage coal-to-shale-gas conversion in the developing world by vigorous sharing of US know-how.

These are all things that any future president needs to know. And they are all things that the future president will need to teach the public, to prevent energy policy from being driven by the public's know'n things that ain't so.

ENERGY
FOR
FUTURE
PRESIDENTS

INTRODUCTION

ENERGY is the most important commodity in the world today. National wealth and energy use are strongly correlated. Countries go to war over energy. We are so sensitive to energy that disruption in a country that supplied only 2% of the world's oil (Libya) triggered a 10% leap in oil prices.

Energy disasters follow one after the other. An enormous oil leak in the Gulf of Mexico threatened to become the greatest ecological disaster in US history. Nuclear accidents, despite optimistic predictions from nuclear supporters, just keep happening—Three Mile Island, Chernobyl, and then, when everything seemed safe again, Fukushima. Now there's a new energy threat: fracking, a natural-gas drilling method that can pollute entire watersheds. And excessive energy use may be leading us into the greatest catastrophe in human history: runaway global warming, accompanied by storms, floods, and—ironically—droughts.

Much of our current economic distress derives from energy. Half of our $50 billion annual balance-of-trade deficit comes from oil imports—and it could get worse. China competes with us for oil, and its oil imports are growing by 50% per year. That heavy reliance on oil puts an enormous stress on the market, particularly since we are predicted to hit (and pass) the oil peak in this decade. Many countries are threatened by energy insecurity. France and Germany discovered their vulnerability when—supposedly to teach Ukraine a lesson—Russia cut off the supply pipelines for a day in 2009.

Despite these problems, we take energy for granted. When electric power to our home is disrupted, someone must be at fault! If prices rise, someone must be cheating! Access to cheap energy in the United States is a necessity, not a luxury. Many people treat it as a fundamental human right.

We hear that we're running out of energy, yet we are inundated with claims that energy is all around, ripe for the picking, a great investment—if only we were smart enough and not being misled by the big energy corporations. There is energy in sunlight, in ocean water, in wind, in the heat that seeps slowly out of the deep Earth. We are told that our dependence on fossil fuels is really just an addiction, manipulated by drug pushers known as oil companies.

We need energy, but we are profligate with the energy we have. Large lecture halls at UC Berkeley (where I teach) use artificial lighting even in midday. We overheat buildings in winter and overcool them in summer, and it feels luxurious to do so. Energy conservation has a bad name, associated with underpowered automobiles, harsh fluorescent lighting, and having to wear sweaters in our own living rooms.

Energy is at the heart of our national security, both military and economic, and it is central to the decisions made not only by presidents but also by citizens. Yet energy is abstract and mysterious. Physics books define it as "the ability to do work"—but that doesn't help unless you understand the technical definitions of *ability* and *work*, and those turn out to be equally abstract. We are told to conserve energy, but physicists tell us that conservation of energy is not a choice but a law of nature. Energy can be very confusing. What does a future president really need to know about it?

When you become president, can't you simply let your secretary of energy or your science advisor handle the energy issues? If only it could be that simple! Imagine the following two plausible scenarios:

- Your science advisor extols the virtues of solar power at the same time that your economic advisor frets about the demise of the US automobile industry and your secretary of state is desperate about a revolution taking place in an oil-rich country like Saudi Arabia or Iran. Knowledge about energy is key for tackling all of these questions.
- Your secretary of energy thinks that the nuclear reactor disaster at Fukushima was so horrible that it should be the death knell of nuclear power, but your science advisor says there is no tsunami danger in the United States, nobody was killed by the radioactivity released, and the accident illustrates the robustness of nuclear power. You have to balance the advice and make the decision, but reconciling the disagreement seems impossible.

How can you reach the right conclusion when your advisors disagree? The answer, of course, is that you have to understand not only their conclusions but also the facts and logic that drew them to those conclusions. We live in a high-tech world, and you can't govern by simply knowing the kind of economic, political, diplomatic, and military information that the presidents of old could use to get by. You need to know and understand energy. And even tougher—you have to lead the public and Congress. You can't operate by taking polls because the public may not appreciate the subtle balance of all the issues. The buck stops with you.

The goal of this book is not to give advice but to educate. I will occasionally give my opinion, but beware—it is the opinion of a scientist, and when you are president I won't be surprised or disappointed if you take a different path. At least it will be an informed path.

I've made this book terse rather than comprehensive. I've emphasized the essentials—the things that the president most needs to know, the fundamental science and information that serve as the entry to deeper understanding.

The book begins with a fresh look at some recent energy disasters. These events make headlines, and those headlines often define the public attitude toward energy. Yet the headlines often turn out to be inaccurate or misleading. In the months and years following a disaster, we often learn that our first impressions were wrong. That is certainly the case with the Fukushima nuclear disaster, with the Gulf of Mexico oil spill, and even with much of the public perception of the effects of fossil fuels on global climate change.

The second part of this book covers the energy "landscape," which is changing with remarkable rapidity. All our modes of transportation—autos, trucks, airplanes—depend on liquid energy (oil, diesel fuel, gasoline), yet we are running perilously low on domestic supplies. There is a windfall in the availability of natural gas—but can we run our cars on it? And how badly will "fracking," the new natural-gas extraction method, damage the environment? How effective will the same technique be when applied to our huge shale oil reserves? Is energy conservation as uncomfortable as many people seem to think, leading to a general decrease in perceived standard of living, or is it something that can be achieved without hardship? Might it actually be a great investment that we would be foolish to ignore?

The third part of the book will review the key "new" technologies, some of them quite old but getting a second breath. How can solar make sense when it goes away at night? Is nuclear power dead, or will it arise again, phoenix-like from the ashes of Fukushima? Whatever happened to the hydrogen economy? Will our children all be driving electric cars? These are issues distorted by headlines in the media and by hyperbole from entrepreneurs.

Near the end, we'll get to a question that, surprisingly, can be deferred: What *is* energy? Physicists have a very good definition, but it is not necessarily helpful to a president who is most concerned with energy policy. So we'll use the immersion method, learning all about energy before we even try to define it. In fact, this part of the book will be optional. You don't have to be able to define energy, as long as you know it when you see it.

Finally, in the last part I'll assume you decided to ask my advice, and I'll give it. This chapter is the least important because my opinions are based on the narrow understanding of a scientist. You'll be the one who has to balance the technological possibilities with the limitations of economics, national security, and international diplomacy.

A good president has to be a leader, and that means more than making the right decisions. The president has to be the nation's instructor. No science advisor or secretary of energy will be able to convince the public that the common perceptions are not necessarily true. That feat can be accomplished only by the person they trust more than any other—the person they elected.

I

ENERGY
CATASTROPHES

ENERGY *is big—the United States consumes 20 million barrels of oil each day—and so energy accidents are often big too. Worse, major energy accidents seem to be happening with increasing frequency: the Fukushima nuclear meltdown in Japan, the Gulf oil spill, the seepage of coal ash waste into the waters of Tennessee, the pollution of streams and rivers by gas well fracking, and possibly the biggest of all, global warming and climate change—perhaps breaching a tipping point that leads us to extremes of climate not yet experienced by civilized humans. These horrors are all consequences of our craving for vast amounts of energy, leading many environmentalists to conclude that the only kind of clean, safe energy is drastically reduced energy.*

Are all energy accidents catastrophes? It certainly seems that way, at least from the first impressions we get from newspaper headlines. Of course, dire voices are the ones most clearly heard. The Gulf oil spill was the greatest environmental catastrophe in the history of the United States—or was it? Fukushima proved once again that nuclear power cannot be controlled—or did it? First impressions are often wrong impres-

sions, and by the time fuller information is obtained, the events have drifted off the front page. So, months or years after a disaster, it is worth taking a new look to see how bad it really was.

Let's reexamine the big three—Fukushima, the Gulf oil spill, and global warming—with the goal of getting past first impressions and looking closely at what we've learned from detailed follow-up investigations. We need to get our facts right, put the consequences in perspective, clear up misimpressions, and get to the core of what really happened, or is still to happen. We can't afford to have our energy policy driven by misunderstanding, confusion, and ignorance.

1

FUKUSHIMA

Meltdown

AN ENORMOUS earthquake struck Japan on March 11, 2011—
magnitude 9.0, thirty times more energetic than the one that
destroyed San Francisco in 1906. Even worse, the earthquake
pounded the ocean and spawned a monster—a 30-foot-tall tsunami,
as high as a three-story building—that struck the coast and smashed
inland, killing more than 15,000 people and destroying over 100,000
buildings.

The most famous victim of the tsunami was Fukushima Dai-ichi,
a nuclear power plant on the coast, placed there to take advantage
of the cooling waters of the sea (Figure I.1). Two workers at the reactor were killed by the earthquake, another by the tsunami, which
is believed to have been 50 feet high at this site. But over the next
few hours and weeks and months, the fear grew that the ultimate
victims of this damaged nuke would number in the thousands or
tens of thousands or more. The power station had been designed to
survive a large earthquake, and it had done that, but nobody had
anticipated a 50-foot tsunami. The reactor was severely damaged.
Might the uranium inside explode like an atomic bomb?

Figure I.1. The Fukushima reactor, in the midst of its meltdown. The steam comes from the water being used to cool the core.

No. Nothing, not a tsunami, not an asteroid impact—not even a terrorist takeover—could cause the Fukushima reactor to explode like a nuclear bomb. The reasons are deeply fundamental, not based on engineering but on the physics of the reactor itself. It takes more than uranium to make a nuclear bomb explode; if that weren't true, many more nations and several terrorist groups would already have such weapons.

Both nuclear bombs and nuclear reactors depend on the nuclear chain reaction, a process in which an atom of light uranium, U-235, explodes—fissions—releasing enormous energy, 20 million times more than the energy released by a molecule of TNT. Fission also emits a few neutrons, small particles that live inside the nucleus. When these neutrons hit other U-235 atoms, they trigger them to fission, making more neutrons. Double the number of neutrons in each stage, and after about 80 stages (taking only a few millionths of a second for all 80 doublings), a pound of uranium will have fissioned, releasing the energy of 20 million pounds—10 kilotons—of

TNT. The fissioned nuclei deposit their energy in the form of heat, and that makes the debris hotter than a thousand suns. The matter is vaporized, ionized, and turned into an enormously high-pressure plasma that blasts out, destroying everything in its way.

For the bomb to work in that way, the uranium must be virtually pure U-235. But in a nuclear reactor, the uranium is typically only 4% U-235, with the rest consisting of heavy uranium, U-238, a form that grabs the neutrons but doesn't fission in a way that can sustain a chain reaction. Because of the polluting U-238, the chain reaction never gets going, unless a trick is employed. The trick, invented by Enrico Fermi during World War II, is to mix carbon or water with the uranium. If there is enough, the neutrons hit these atoms first and lose energy before they collide with U-238. The neutrons are slowed—that is, "moderated" by the collisions. A peculiar but important feature of U-238 is that it doesn't readily absorb such slow neutrons; they tend just to bounce off. Eventually the neutron will hit U-235, and the chain reaction will continue. A nuclear reactor is designed so that, on average, only one of the emitted neutrons goes on to trigger a new fission, so the rate of energy release remains steady.

The required slowness of the neutrons is what prevents a big explosion. If something goes wrong and the chain reaction begins to run away, we have what is called a *reactivity accident*. The energy builds up, but because the neutrons are moving so slowly, the explosion develops slowly too. When the energy density reaches the level of TNT, the reactor blows apart, stopping any further chain reaction. The energy released is comparable to that of TNT, 20 million times weaker than the energy released by an atomic bomb.

In 1986, the Chernobyl nuclear reactor did explode, dynamite-like, from a runaway nuclear chain reaction—a reactivity accident. Look at the photo in Figure I.2. The explosion was enough to destroy most of the reactor building, but that was it. The disaster that followed didn't come from the explosion, but from the enormous amount of radioactive debris that was released. The number of cancers caused by radiation released is plausibly estimated at

Figure I.2. The destroyed Chernobyl plant. Although it was a runaway chain reaction, the 1986 Chernobyl explosion was too small to destroy much more than the reactor building that housed it.

24,000; fortunately, many of these are thyroid cancers and are readily treatable.

Unlike at Chernobyl, the reactor at Fukushima did not explode. A buildup of hydrogen gas caused the upper building to explode, but the reactor itself survived the tsunami, turned itself off, and sat safely for several hours. Even though the chain reaction had stopped, there was enough radioactive material in the core to generate dangerous levels of heat, but cooling pumps initially kept the situation from getting out of hand. The most modern reactors don't need these pumps; they are designed to use the natural convection of the water to keep it circulating. But the Fukushima reactor was not the most modern kind. It depended on auxiliary power systems to keep the pumps going. These systems worked well, despite the devastation of the earthquake and the tsunami, and they kept the reactor cool.

Of course, this auxiliary cooling wasn't expected to work forever; it was designed to last about 8 hours, after which normal power was supposed to be restored. That plan did not anticipate the enormous destruction of the infrastructure that was wreaked by the tsunami. The emergency power ran out and the fuel overheated, much of it melting. Technically, the Fukushima accident was called a *station blackout failure*, since it was the loss of electricity that killed the plant. The meltdown led to a huge and dreaded release of radioactivity—a release much greater than had happened in the prior US nuclear reactor accident at Three Mile Island in 1979. In fact, the enormous release of radioactivity was second only to that of the 1986 Chernobyl nuclear reactor accident.

Radioactive Release

When the uncooled fuel in the Fukushima nuclear reactors overheated, it melted the metal capsules that contained the fuel and the nuclear waste. Volatile gases spewed out, and then much more. Most dangerous are the radioactive forms of iodine and cesium: I-131 and Cs-137. Iodine is bad because its atoms have a short half-life; it decays rapidly, releasing radiation as it goes—thus making it the biggest initial source of radioactivity. Moreover, if you inhale or consume iodine, it concentrates in the body in the thyroid, where it can induce cancer. Ironically, the fact that iodine decays rapidly is also good news. Half of it is gone in 8 days, and in just 2 months the radioactivity of I-131 drops to about one-half of 1% of its initial level. By now, all the I-131 released from Fukushima is gone.

There is a protective measure you can take if you're near an I-131 release: swallow some potassium iodide pills. Your thyroid will become saturated with (nonradioactive) iodine and won't absorb any more. You are temporarily protected from I-131. Wait a few months, and your thyroid is safe. Nevertheless, the dose that many people breathed in following the Fukushima accident was large.

Radioactivity from cesium is initially lower because it decays more slowly, but that means the radioactivity lasts longer. It takes 30 years for half of the cesium to go away. Radioactive strontium (Sr-90) is equally slow. These elements expose you to radiation more slowly, but the slow pace of radiation release means they hang around longer, making them more insidious. They can settle on plants and be

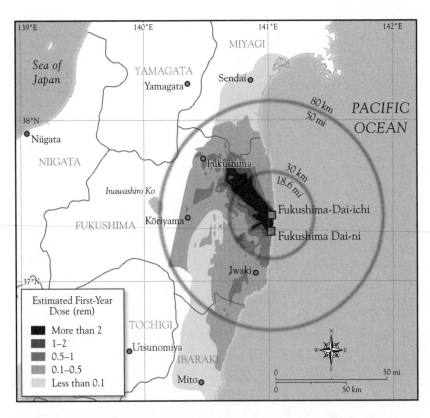

Figure I.3. Map of the Fukushima region, showing the expected dose for each location that would have been received by any person who remained there for a year after the accident. The darkest region, running from the coast to the northwest, is marked as greater than 2 rem. Distant regions are below 0.1 rem.

eaten by animals and humans, where they become concentrated in bones.

One horrible fact of the 1986 Chernobyl disaster was that the public was not protected from the iodine and cesium and strontium until a great deal of harm had been done; residents continued to drink milk from cows that had eaten contaminated grass. In contrast, officials at Fukushima ordered evacuation of the accident vicinity and banned the consumption of food grown in the region.

How bad was the Fukushima contamination? The map in Figure I.3 shows the radioactive dose for a hypothetical resident who remained at that location from the day of the release until a year later. It does not include radioactivity from food grown or raised in the region, since such food had been banned.

Radioactivity and Death

Radiation damage to the body is measured in units called "rem." If you are exposed to a dose of 100 rem or more, you'll get sick right away. It's called *radiation illness*. You know what that's like if you or a friend has undergone radiation therapy: nausea, loss of hair, general feeling of weakness. In the Fukushima accident, nobody got a dose this big; workers were restricted in their hours of exposure to try to make sure that none received a dose greater than 25 rem (although some exceeded this level). At a larger dose—250–350 rem—the symptoms become so severe that they threaten your life; essential enzymes are damaged, and your chance of dying (if untreated) is 50%.

Nevertheless, even a small number of rem can trigger an eventual cancer. A dose of 25 rem causes no radiation illness, but it gives you a 1% chance of getting cancer—in addition to the 20% chance you already suffer from "natural" (not understood) causes. For larger doses, the danger is proportional to the dose, so a 50-rem dose gives you a 2% chance of getting cancer; 75 rem ups that to 3%. The cancer effects of these doses, from 25 to 75 rem, are well established by studies of the excess cancers from the Hiroshima and Nagasaki exposures. We don't know if the proportionality continues to hold at very low doses, but for the moment, let's assume it does.

Here's another helpful way to think about the calculation. If 25 rem gives you a 1% chance, then a cancer *dose* can be defined to be 2,500 rem (25 rem × 100). Of course, a dose that high would kill you from radiation illness. But if spread out over 1,000 people, so that every person received on average 2.5 rem, then the 2,500 rem would still induce one extra cancer because even if shared, the total number of cells damaged by the same number of nuclear decays would be the same. In fact, if spread out over a million people, 2,500 rem would still induce only one cancer (on average). The dose measures radiation damage, and if there is one cancer's worth of damage, it doesn't matter how many people share that risk.

I can summarize the calculation with a simple formula. If you want to know how many excess cancers there will be, multiply the population by the average dose per person to get the *person-rem*, and then divide by 2,500. We can use this method to calculate the Fukushima cancers. Let's take a closer look at the radiation doses for that accident.

The dark swatch running inland from Fukushima in Figure I.3 looks ominous. The radiation dose for that innermost region is 2 rem on the boundary, and higher than 2 rem in the interior. A particularly high dose, 22 rem, was measured in a section of the town of Namie, located 14 miles from the reactor. This value represents the total dose up to evacuation, which for Namie was on April 22. Afterward, the high levels of radiation dropped quickly; the largest component came from iodine, and its level dropped by 50% every 8 days.

How many cancers will such a dose trigger? To calculate an answer, we'll assume that the entire population of the innermost region, about 22,000 people, received the highest dose: 22 rem. (This *conservative* assumption invariably overestimates the danger.) The number of excess cancers expected is the dose (22 rem) multiplied by the population (22,000), divided by 2,500, equals 194 excess cancers. Those cancers are the tragedy of the relatively late evacuation.

Let's compare that to the number of normal cancers in the same group. Even without the accident, the cancer rate is about 20% of the population, or 4,400 cancers. Can the additional 194 be detected? Yes, because many of them will be thyroid cancer—a kind that is normally rare (but treatable). Other kinds of cancers will probably not be observable, because of natural statistical variations of normal cancers.[1] Sadly, many of those 4,400 who die from "normal" cancer will die believing their illness was caused by the nuclear reactor. That's just human nature; we search for reasons behind our tragedies. The causes of normal cancer are not understood, although smoking is a known contributor. Of the roughly 100,000 survivors of the Hiroshima and Nagasaki blasts, we expect about 20,000 have died or will die from cancer; however, only about 800 of them had their cancer caused by the bombs. We know that by looking at similar cities: there is an increase in cancer among those exposed, but it is only a small increment of the natural rate. Yet far more than the estimated 800 victims attribute their cancers to the bomb.

What about the Fukushima outlying regions? The next zone on the map had a population of about 40,000 and an average dose of 1.5 rem, yielding a total dose of $40,000 \times 1.5 = 60,000$ person-rem, making the number of expected extra cancers $60,000 \div 2,500 = 24$.

Do these numbers seem low to you? They are tragic, as are all deaths, but smaller than the impression that people get from newspaper headlines. Thanks to the evacuation, the total number of deaths the radioactive release has induced in and near Namie will almost certainly be less than 300, and a more reasonable estimate (using average exposures rather than the maximum ones) suggests that the number will be less than 100. One hundred cancer deaths is bad, but that number is minuscule compared to the 15,000 deaths caused by the tsunami.

What about more distant regions? Even a tiny bit of radiation averaged over a huge population could conceivably cause cancer. True, but we are immersed in "natural" radioactivity from cosmic rays (radiation coming from space) and from the earth (uranium, thorium, and naturally radioactive potassium in the ground). These natural levels are typically 0.3 rem per year; plus we are exposed to an additional 0.3 rem if we include average medical exposures from X-rays and other medical treatments. Some areas have even higher natural levels.

The Denver Dose

Denver, Colorado, has particularly high natural radioactivity, coming primarily from radioactive radon gas, emitted from tiny concentrations of uranium found in local granite. (Some people mistakenly believe that most of the excess radiation comes from stronger cosmic rays at the high Denver altitude.) If you live in Denver County you get an *extra* dose of 0.3 rem (300 millirem) per year, on average. Radon doses for more US locations can be found at Radon. com (http://www.radon.com/radon/radon_map.html). Look up your home region. The site gives exposures in picocuries per liter (pCi/l); to convert to rem per year, multiply by 0.09.[2] You'll often come across another unit, called the *millisievert*. One millisievert is 0.1 rem = 100 millirem. Thus, the excess dose in Denver is 3 millisieverts per year.

It is worth noting that despite its high radiation levels, Denver generally has a lower cancer rate than does the rest of the United States. Some scientists interpret this as evidence that low levels of radiation induce cancer resistance; I think it more likely that lifestyle differences account for the disparity.

Let's continue with our assumption that low doses of radiation are proportionally bad for your health. By our cancer dose calculations, the excess Denver dose of 0.3 rem means that the chance of dying from a radiation-induced cancer is $0.3 \div 2{,}500 = 0.00012$. This probability is so tiny that it is impossible to measure or detect in human populations.[3] Should an undetectable danger play a major role in determining policy? That is not a trivial question. It is one that a future president needs to think about.

On October 15, 2011, the *New York Times* ran a front-page story titled, "Citizens' Testing Finds 20 Radioactive Hot Spots around Tokyo." If you read on to page 3, the article states that the hot spots that frightened everyone showed radiation at the level of 1 milli-sievert per year; that's 100 millirem, or 0.1 rem. This, the report stated, was the same level that had led to continued mandatory evacuation at Chernobyl. The article didn't mention that the number is small compared to the average excess dose that people happily live with in Denver.

Why was the evacuation level for Chernobyl set so low? Some people advocate use of the *precautionary principle*—the idea that when there is uncertainty, you should consciously err on the side of safety. The International Commission on Radiological Protection (ICRP) recommends evacuation whenever the radiation dose exceeds 0.1 rem per year; yet that's one-third of a Denver dose! Is the commission's recommendation erring on the side of caution? Not necessarily. It could be argued that the disruption to lives caused by the continued Chernobyl evacuation at 0.1 rem (and the fright in Tokyo) did more harm than would have been done by the radiation itself. The side effects of the prescription may have been worse than the illness it treated. Applied strictly, the ICRP standard might require the immediate evacuation of Denver.

After the 1979 Three Mile Island nuclear reactor accident in Pennsylvania, the Kemeny Commission was convened to study the health damage of the released radiation. The commission concluded that the primary damage was not cancer, but psychological stress induced by unnecessary panic. It is likely that more people were harmed by stress-induced cigarette smoking than by any actual reactor release.

One of the oddest facts is that some newspapers in Denver warned of the radioactive cloud from Fukushima that drifted all the way across the ocean. They did that even though the observed radioactive cloud was measured in microrem—a thousand times weaker than the natural excess radioactivity in the area. Maybe the alarm was a consequence of the mistaken belief that reactor radioactivity is somehow more dangerous than "natural" radioactivity. Or maybe most people just don't know that they already live in a naturally radioactive world.

Read the newspapers and you'll sometimes see numbers for expected cancer deaths from Fukushima that are higher than the ones I've estimated here. The most thoughtful high-number calculation comes from Dick Garwin, a renowned nuclear expert. He has written that the best estimate for the number of deaths is about 1,500—well above my number of 100 (but still only 10% of the immediate tsunami deaths). Garwin uses the same numbers that I use, but he extrapolates forward in time 70 years to the continuing damage that residual radiation could cause, assuming that the radiation cannot be covered, cleaned, or washed away, and that the population of Fukushima doesn't change. Moreover, he ignores the Denver Dose argument and includes in the calculation the numbers of deaths expected from tiny doses, assuming that even small exposures are proportionately dangerous. (This is an assumption that has also been adopted by the prestigious US National Academy of Sciences.) I don't dispute Garwin's number, but I believe it has to be understood in context. If you apply the same approach to Denver, you have to take into account the fact that the Denver Dose is delivered every year. Then, over 70 years, it sums to 0.3 rem × 70, or 21 rem per person. Multiply that by 600,000 people (the current population of Denver) and divide by the cancer dose of 2,500 rem to get the expected cancer excess in Denver of 5,000, over three times higher than Garwin's Fukushima number.

I am uncomfortable with both of these large numbers. They are based on a theory of cancer that predicts proportionality—but it is a theory that has never been tested, will not be tested in the foreseeable future, and which is known to fail for leukemia. At the same time I can't be sure that the theory is wrong. But I consider both of these relatively large numbers to be misleading. Remember that Denver has a lower cancer rate than the rest of the United States, not higher. The value of ignoring dangers below the Denver Dose is that in doing so we are ignoring risks that are unobservable and which we routinely ignore (and properly so) in other circumstances.

Even though Garwin prefers the 1,500 number, he also says that it is small enough that evacuation of Fukushima would probably cause more harm than good. Evacuation causes disruption to lives that is hard to quantify but is very real.

Some people believe that the untested proportionality assumption should be made because it gives you a "conservative" estimate. Beware of that adjective. What is conservative depends on your agenda. Is a conservative estimate one that likely overestimates deaths? If so, then it is likely to lead to more disruption through evacuation and panic; is that truly conservative? In this book I'll try to give "best" estimates, rather than conservative ones, and let you adjust them according to what you consider most important. I recommend that when you are president, you demand best estimates from your cabinet members and advisors, not conservative ones; otherwise, your numbers will incorporate the biases of the person who gave them to you.

Another way to overestimate the deaths is to use a much higher value for the induced cancer risk than has been determined by the best scientific studies. For example, the movie *Chernobyl Heart* attributes many deaths and illnesses in the Chernobyl region to the release of radioactivity, whereas careful analysis attributes the problems to the high cigarette and alcohol consumption that has long been endemic in the region.

I think the most useful estimate is the one I've given: from the radiation so far, fewer than 100 induced cancers. Residents of Fukushima who are concerned that residual radiation will cause additional risk can avoid that by leaving, but they need to recognize that any additional cancers will be statistically unobservable, hidden well below those of natural cancer and the other dangers of modern life.

Bottom Line: What Should We Do?

The tsunami was horrendous. Over 15,000 dead from the giant wave itself. It also caused a severe meltdown in a nuclear reactor, leading to cancer deaths and additional consequences for the surrounding region, including the evacuation of populated areas not hit by the wave. The economic consequences of the reactor destruction were terrific. The human consequences, in terms of death and evacuation, were also large. But the radiation deaths will likely be less than 100, a number so small compared to the tsunami deaths that, tragic as they are, they should not be a central consideration in policy decisions.

The reactor at Fukushima was not designed to withstand a 9.0 earthquake or a 50-foot tsunami. Surrounding land was contaminated, and it will take years to recover. But it is remarkable how small the nuclear damage is compared to that of the earthquake and tsunami. The backup systems of the nuclear reactors in Japan (and in the United States) should be bolstered to make sure this never happens again. We should always learn from tragedy. But should the Fukushima accident be used as reason for ending nuclear power?

Nothing can be made absolutely safe. Must we design nuclear reactors to withstand everything imaginable? What about an asteroid or comet impact? Or a large-scale nuclear war? No, of course not; the damage from the asteroid or the war would far exceed the tiny added damage from the radioactivity released by a damaged nuclear power plant.

It is remarkable that so much attention has been given to the Fukushima radioactivity release, considering that the direct death and destruction from the tsunami crashing into the coast was 100 times greater. Perhaps the reason for the focus on the reactor meltdown is that it's a solvable problem; in contrast, there is no plausible way to protect Japan from 50-foot tsunamis. Do we order a permanent evacuation of the coast to 20 miles inland? Do we try to build a 50-foot-high seawall all around the eastern coast, including Tokyo Bay?

Here's the guideline I propose for nuclear power plant safety: Make a nuclear plant strong enough that if it is destroyed or damaged, the incremental harm done by the release of radioactivity is small compared to the damage done by the root cause. If you have extra money to spend, spend it on avoiding the causative disaster, not the secondary one.

In addition, for radioactivity concerns, I suggest that the Denver dose be adopted as a standard. In planning and in disaster response, completely ignore any level of radioactivity that is less than the excess naturally received every year by the residents of Denver: 0.3 rem = 300 millirem = 3 millisieverts. The ICRP evacuation dose should be raised at least to this level. And recognize that even a dozen Denver doses might do much less harm than an evacuation or otherwise exaggerated response.

By these criteria, the Fukushima reactor complex was adequately designed. New reactors can be made even safer, of course, but the bottom line is that Fukushima passes.

The great tragedy of the Fukushima meltdown is that as this is being written (early 2012), Japan is shutting down all its nuclear reactors. The hardships and economic disruptions induced by this policy will be enormous, and dwarf any danger from the reactors. Perhaps taking action on something nuclear helps deflect attention from the true danger in Japan, one that they have no protection against—the threat of another giant earthquake and tsunami.

2

THE GULF OIL SPILL

PRESIDENT Obama called it "the greatest environmental disaster of its kind in our history." He went on to say that this catastrophe was destroying or threatening "not just fishable waters, but a national treasure"—the beautiful and bountiful Gulf of Mexico and its surrounding shores. "Already, this oil spill is the worst environmental disaster America has ever faced. And unlike an earthquake or a hurricane, it is not a single event that does its damage in a matter of minutes or days. The millions of gallons of oil that have spilled into the Gulf of Mexico are more like an epidemic, one that we will be fighting for months and even years."

On July 15, 2010, after nearly 3 months of effort, the flow was finally stopped. On that day, presidential advisor David Axelrod topped Obama's depiction, saying that the Gulf oil spill was the "greatest environmental catastrophe of all time."[4] That description classified the Gulf spill as worse than not only the terrible dust bowls of the 1930s, but also even historical disasters such as the destruction of the forests of Europe in the Middle Ages. The hyperbole was growing, and when that happens, it's worthwhile to look hard at the data.

In the end, over 6,000 dead animals were found, mostly birds.

A particularly heart-wrenching photograph of an oil-soaked pelican was widely published (Figure I.4). To put the number of animal deaths in perspective, note that the US Fish and Wildlife Service estimates that glass windows in buildings in the United States kill between 100 million and 1 billion birds each year. Collisions with high-voltage electric lines kill another 100 million. And tens to hundreds of thousands of birds are killed every year when they get trapped in fishing nets.

After the *Exxon Valdez* shipwreck in 1989, there were many ugly images of formerly beautiful beaches covered with tar. For the Gulf spill, similar beach photos were missing. On the TV news I watched video footage of a new cleaning vehicle, a monstrous tractor that scoured the beaches to clean up the mess. I expected images of oil and tar and gunk being picked up in an attempt to return the beaches to their pristine state. But the news media never focused on

Figure I.4. Soon after the Gulf oil spill, one of the most distressing photos was this one of a pelican. It almost seemed to be crying out, "Why did you do this to me? How could you let this happen?"

the sand, only on the machines. As best I could tell, studying the background in the film, there was little tar on the beach. I became somewhat cynical. The company behind the spill was BP, the huge conglomerate once named "British Petroleum." Had the local community convinced BP to pay for a beach cleaner that it wanted for other reasons? (When I was in ninth grade, I had a summer job as a beach cleaner, and it is amazing how much clutter has to be removed between six and nine o'clock every morning from the messy beach uses of the prior day.)

It turns out that BP and other oil companies had learned one lesson very well from the earlier spills: how to keep the oil from reaching the beaches. BP hired local fishermen (whose normal livelihood had been interrupted anyway) and implemented an extensive plan of distributing buoys and barriers, and of dumping tons of dispersants—the equivalent of soap—to make the oil break up and dissolve in the water. The result was enormously successful. Virtually all of the beaches remained clean.

That was good news. Tourism could rebuild. But there was lots of potential damage that would not be so easy to photograph for You-Tube. What about out at sea? What would be the effect on fishing? And what would be the consequences for the coming decades?

In events such as this, it is difficult to separate truth from hyperbole. When you hear that a particular event is "the greatest catastrophe of all time" but don't see horrific images in the media (or see the same one over and over, as with the pelican photo), then there are typically three plausible scenarios:

1. *The damage will eventually become evident.* For this spill, we would soon see dead porpoises and whales, and eventually extensive tar and other indicators of a deep catastrophe. Tourism would be decimated for years.
2. *The damage is hidden.* Even though things looked good on the Gulf beaches, surveys and scientific investigations would show conclusively that the deep ocean had been severely damaged. The remoteness of this wilderness does not make it less important.
3. *The accident is not a catastrophe but is nevertheless described as a "near*

catastrophe," something that could have been much, much worse— if, for example, BP had not been lucky in capping off the well, or a hurricane had hit at just the wrong moment.

Now that some time has passed, the Gulf accident has moved solidly into the third category. Fishing has resumed, and tourists are going back to the still-beautiful beaches on the Gulf Coast. True, there are some claims for the second scenario—hidden damage— but that scenario tends to be speculative and somewhat exaggerated. Unfortunately, it is not in the short-term interests of either the public or the government to conclude that the event was less than a catastrophe. The public, even the part of it that was affected only indirectly (for example, by a decrease in business from reduced tourism), wants to be reimbursed for lost revenues. The government does not want to admit that it exaggerated the disaster and perhaps, through its severe measures, actually caused much of the distress to the public itself. And the government, too, is looking with anticipation at the billions of dollars it could extract from BP.

Note that the list of scenarios doesn't include the possibility that the threat was exaggerated, that the accident wasn't even a near catastrophe, that it was bad (it killed 11 people) but environmentally tolerable—something with no lasting consequences and easily prevented in the future by a few regulatory adjustments. This fourth scenario is not politically attractive, since the government and the media don't want to admit that they overreacted; and it is even denied to the culprit (BP in this case), since it would be seen as a self-serving attempt to escape liability.

To understand these issues, it is helpful to look back in more detail at what happened.

The Deepwater Horizon Accident

It began on April 20, 2010, when an explosion on the *Deepwater Horizon* oil rig in the Gulf of Mexico exploded, killing 11 workers and injuring 17 more. The rig had an appropriate name: 41 miles out at sea, even with its height of 320 feet, it was invisible—below the horizon—from the Louisiana coast. This rig was not sitting on the seabed; it was floating on 5,000 feet of water, with a long flexible pipe connecting it to the oil source, an additional 18,000 feet below the seafloor. Because of the weight of the rock above it, the oil was under tremendous pressure. When the rig exploded, the pipe was ripped open and oil gushed out—at a rate of 26 gallons per second.[5]

Imagine the movies of the old oil well gushers—when a drill hits a highly pressurized oil pocket. Oil shoots upward, uncontrolled, spewing over the land and (at least in the movies) the delighted owners of the rig, who now know they are suddenly rich. In real life, such gushers are considered a terrible waste of oil, and riggers work hard and fast to close off the pipe. That's difficult enough to do on land, but the *Deepwater Horizon* site was under nearly a vertical *mile* of salt water. Remote vehicles could swim down and take photos, but doing more was difficult. Capping a gusher is tricky under the best of circumstances.

Millions of gallons of oil poured into the sea. Every day the *New York Times* published a map illustrating the extent of the oil spill. Figure I.5 shows the map on June 26, 2010, just over 2 months after the accident. The large polygon shows the area that President Obama closed to fishing, extending 400 miles off the coast.

There were dire predictions: The west coast of Florida was likely to be ruined. Oceanographers expressed the possibility—even the likelihood—that the spill would round the peninsula of Florida, flow with the Gulf Stream, and pollute the waters and the shores of the eastern United States, perhaps as far north as Cape Cod and Maine, and maybe even Europe.

Yet as I looked at the daily maps, I saw that the surface oil did not seem to be spreading as feared. The map in Figure I.5 shows the approximate *maximum* extent of the oil. The oil continued to flow (it was not plugged until July 15), but the size of the slick was not increasing. It never got even close to Florida. Where was the oil going? We think we know the answer, at least partially. Some was evaporating, some was dispersing in water, some was sinking, and some was removed by the cleanup effort.

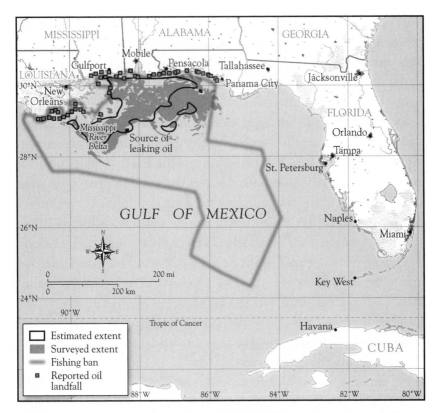

Figure I.5. The maximum surface extent of the Deepwater Horizon *oil spill, from the* New York Times, June 26, 2010.

Crude oil consists of a mixture of chemicals called *hydrocarbons*. A hydrocarbon molecule is a chain of carbon atoms surrounded by a sheath of hydrogen atoms. The shortest chains are methane with 1 carbon, ethane with 2, propane with 3, and butane with 4. The shortest hydrocarbons are gases, and they disappear into the atmosphere. Somewhat heavier oil molecules (octane, with 8 carbons; and cetane, with 16) are light liquids, and they float; that's what you see in the satellite images of the Gulf spill. Just like ordinary gasoline (which is a mixture of such hydrocarbons), these eventually evaporate.

If the surface layer is thick enough, the oil can be burned. Thin layers eventually disperse or dissolve in the water, making the water potentially toxic until natural circulation dilutes it sufficiently.

The entire spill is estimated to be about 250 million gallons. That's about a million cubic meters—about a million cubic yards, a cubic football field. (For the purposes of this kind of estimate, a meter is equal to a yard, to sufficient accuracy.) That's a lot—but the ocean is big. If the entire spill, rather than being dispersed or evaporated or sunk, filled the dark "surveyed extent" area in Fig. I.5 (about 10,000 square miles), then the depth of the oil layer would be about two-thousandths of an inch thick, about as thick as a typical piece of paper or a thick human hair. That is thin enough that if it were dissolved only in the water directly beneath it—5,000 feet's worth— the oil would be diluted to less than one part per million—below what is considered the toxic level. The spill was big—and we don't want to have more such spills occur—and it did harm wildlife, but this is nevertheless an example in which the true solution to pollution really is dilution. There is a lot of ocean out there.

The purpose of the much criticized surfactants added to the water was to make sure that the oil did not clump into bigger blobs— simply because the bacteria that eat the blobs can contact only their outer surfaces, so big blobs disappear slowly. And it is the big blobs that can gum up the feathers of seabirds and the fur of sea mammals.

The much heavier molecules (containing 20 or more carbons) coagulate into tars, and those with a density greater than water sink. Underwater, these tars are attacked by oil-devouring bacteria. How quickly? We don't really know, because the deep ocean is exceedingly difficult to observe. Most of the claims of lingering ecological disaster following the Gulf oil spill are based on the assumption that the bacteria act slowly, and that the sunken oil is killing life on the seabed.

The ability of these bacteria to consume sunken oil came as a surprise to many, but not to the experts. The Gulf had recovered rapidly from other spills, including the Ixtoc disaster of 1979. There are many natural leaks in the Gulf, although the yearly sum of them all is probably less than 1% of the *Deepwater Horizon* spill. Nevertheless, that natural supply was sufficient to let oil-consuming bacteria thrive in the Gulf. For them, the day the *Deepwater Horizon* blew was the beginning of a great feast.

The flow was finally stopped on July 15, 2010. Two months later, on September 19, the well was officially sealed.

Damage

How much damage was truly done? As I mentioned earlier, 11 people were killed in the explosion of the *Deepwater Horizon* rig, and an additional 17 were injured.

The spill was in a region of the Gulf that oceanographers call the "blue ocean." William Nierenberg, former director of the Scripps Institution of Oceanography, once explained to me that this part of the ocean is considered a *desert* by oceanographers. It is visited by passing sharks but is so low in nutrients that it is not truly a source of life. To an ecologist, he explained, *desert* does not mean hot and dry; it means low density of life. Closer to shore than the blue ocean is the "green ocean" where nutrients from upwelling and rivers (such as the Mississippi) provide the basis for abundant life. An oil spill in deep water is harder to stop, but it is also more remote from the truly productive, life-giving, nutrient-rich green-water regions.

After the well was sealed and most of the oil on the surface was gone, BP reported a tally of killed wildlife found: 6,814 animals—6,104 birds, 609 sea turtles, 100 dolphins and other mammals, and 1 other reptile. Most of these were probably collected at sea, and BP clearly did not encourage newspaper photos of them. Of course, it is hard to know how many they missed, but lots of people were searching. It was much more than one photogenic pelican. The bluefin tuna may have been particularly hard hit; the spill occurred in their spawning ground during the spawning season, and the Ocean Foundation estimated, using satellite images, that the spill may have reduced the number of bluefin juveniles by over 20%.

But how big were the environmental consequences really? This question is still highly disputed. Like the cancers caused by Fukushima, most of the disaster is calculated, not observed, and those calculations are highly uncertain. Some people make "conservative" calculations but, as with Fukushima, the meaning of *conserva-*

tive depends on your outlook. Do you want to make sure you don't underestimate the damage (a point of view taken by some environmentalists), or do you want to make sure you don't overestimate it because of the damage that an inflated estimate could do to fishing, tourism, and industry?

Some people think that the greatest damage of the accident came from governmental overreaction. After the spill, President Obama declared a moratorium on deepwater drilling, and that ban had a serious impact on the local economy, not only in the thousands of jobs temporarily put on hold, but also in the oil income of the region. Of course, it is tricky to second-guess such a decision. If there had been a second spill, Obama would never have been forgiven. He was undoubtedly acutely aware of the long legacy of problems that mishandling Hurricane Katrina had created for his predecessor, George W. Bush.

Huge areas were designated off-limits for fishing, and the initial worry was that the industry would be devastated. Many of the fishermen were hired by BP to help install the barriers that stopped most of the oil from reaching the shores. Of course, most of them would much rather have been fishing. Cleaning oil is not only dirty work, but it has a potential danger from oil fumes. In some sense, the prohibition on fishing gave BP exactly the kind of experienced boatmen, knowledgeable about local waters, that the company needed to do a good job.

The waters of the Gulf were, region by region, gradually reopened to recreational and commercial fishing of fish, shrimp, oysters, and crabs. The entire region, including the 1,000 square miles closest to the wellhead, was finally opened a year after the initial spill. By then, testing showed that 99% of the water samples had no detectable residues of oil or dispersant even 1,000 times lower than the federal limits. Fish and shrimp were thriving; fishing was excellent. It was as if the wildlife had benefited more from the fishing moratorium than they had been hurt by the pollution. It was truly amazing how well the Gulf recovered—at least the parts that we could measure.

The tourism industry sustained serious but temporary damage. It wasn't just hotels that suffered, but also local restaurants and other recreational facilities. According to the US Travel Association, the disaster affected over 400,000 travel industry jobs, and the negative impact on tourism in the region "could" exceed $23 billion over the ensuing 3 years. It's always hard to know how to interpret numbers such as these, since in this case the group reporting them has a vested interest in suing BP. The size of the number depends on how rapidly the region returns to its previous vigorous tourism.

Iatrogenic Disease

Who was to blame for the *Deepwater Horizon* disaster? Most people think the answer is obvious: BP, the owner of the oil lease. Indeed, BP immediately accepted responsibility. Soon after the spill, the company announced (its spokesperson was President Obama) a trust fund of $20 billion to compensate victims of the disaster. To some, that seemed like a lot, but without knowing how long the damage would continue, others termed it grossly inadequate.

Of course, the spill exacted enormous damage to BP itself. First, there was the $11.2 billion that BP spent sealing the well. In addition, about $40 billion worth of lawsuits have been filed against the company. BP, in turn, is suing three other companies: Transocean, the owner and operator of the *Deepwater Horizon* oil rig (which had cost $500 million to build); Cameron, the manufacturer of the "fail-safe" blowout preventer; and Halliburton, who did the well cementing.

The biggest damage to BP may have been to its name. A few years earlier it had changed its name from British Petroleum to BP, kidded in its ads that the letters referred to "Beyond Petroleum," and worked hard to build a reputation as an environmentally aware and sensitive energy company. Following the *Deepwater Horizon* disaster, people across the country started boycotting BP, and some gasoline stations that carry the BP logo joked that they should change their name back to Amoco—the US brand that BP used until the late 1990s.

The 6-month moratorium on oil drilling off the Gulf Coast represents, according to economists, several billion dollars in economic losses.[6] In addition, the drop in stock capitalization was estimated to be more than $100 billion, but that's possibly a temporary overreaction.

In retrospect, it is proper to ask how much damage was avoided by the swift and intense preventive measures, and how much damage was done by those preventive measures themselves. In medicine, the term *iatrogenic illness* refers to sickness you contract because of a visit to the doctor or because of a procedure used that proved to be more harmful than helpful. An example of an iatrogenic illness is a disease contracted in the waiting room from another patient. Iatrogenic deaths occur in operating rooms from misapplication of anesthetics and other mistakes in procedure. How much of the *Deepwater Horizon* disaster was iatrogenic?

When the *Deepwater Horizon* disaster began, the government got deeply involved in technological methods to stop the flow. It was obvious, however, that the government lacked any useful technology to supplement or replace the efforts of the deepwater industry. But did government hyperbole exacerbate the problem? The beaches remained clean. Was the tourism loss caused only by the spill itself, or was some caused by the exaggerated statements of President Obama and his staff?

Just one year after the spill, Louisiana Governor Bobby Jindal pronounced the region reborn. Local media celebrated the recovery by eating Gulf shrimp. Edward Overton, an ecologist and professor at Louisiana State University, said the environmental effects were far less serious than those of the 1989 *Exxon Valdez* oil spill in Alaska. "We're way off what *Exxon Valdez* was, way off," he said.[7]

In the end, the Gulf spill was not so much a great catastrophe as it was a *near* great catastrophe, something in the third category of disaster scenarios, something that could have been much, much worse—if, for example, BP had not been able to cap off the well or a hurricane had hit at just the wrong moment. Of course, we don't really know *how near* it was to something much worse. Were we lucky—or unlucky?

What does the Gulf oil spill experience mean for energy policy? It is clearly tempting for political leaders to try to be "conservative" and operate assuming the worst-case scenario. They adopt the precautionary principle and try to err on the side of safety. But there is a potential bias in doing that—to take the worst case as the one that is most evident to the public through the media. In fact, harm to a local economy or even the national economy can turn out to be the greater threat. The Gulf oil spill was harmful to the environment, but it is my sense that the overreaction to the spill was even more damaging.

That judgment depends, of course, on the relative weight you give to the environment and the economy, and reasonable people can (and do) differ vehemently on the answer to that balance. The meaning of precaution is in the mind of the advocate, and for that reason, as a simple guideline, the precautionary principle doesn't work. Passions can be so intense that people often choose out of conviction rather than by making a balanced judgment. If the exaggerated view persists in the public image, then it can continue to bias policy in a detrimental way. Should we stop drilling in the Gulf and other waters offshore? Maybe. But we should not base our decisions on the hyperbole of the Gulf spill. When you are president, it will be your job to make sure the decision balances all factors. Doing that wisely, of course, might make you vulnerable to a populist attack that could hurt your chances for reelection.

3

GLOBAL WARMING
AND CLIMATE CHANGE

THE GREATEST energy catastrophe of them all might be global warming and the closely linked threat of climate change. Many people consider these to be the most important issues of our time. They worry that the current generation is leaving a terrible legacy for the future, that our children and our children's children will suffer the consequences of our inaction. Other people believe that climate change is the biggest scam of all time, driven by politicians who want to scare you and scientists who love media attention and increased funding.

It is exceedingly difficult to eke out the truth, in large part because newspapers report the extreme viewpoints and present the subject as a great debate rather than a scientific discussion. Each extreme accuses the other of spewing nonsense, and they are both right. The truth lies deeply buried in the middle.

Here is my executive summary of what a future president needs to know about global warming and climate change:

1. Most of the evidence, as presented to the public, is exaggerated or distorted.

2. Global warming is indeed real and dangerous, and it is worthy of serious effort to stop.
3. Assuming the theory is correct (it may not be), none of the well-known proposals to stop global warming that have been made have any realistic chance of working, even if they are fully implemented.

All three of those statements might seem surprising, or even wrong, so after a brief introduction I'll go over the details that lead me to these conclusions.

A Brief Introduction to Global Warming

Over the past century, human use of fossil fuels has increased so much that the levels of carbon dioxide (CO_2) in the atmosphere have gone up about 40%. Carbon dioxide is still a minor gas—only about 0.04% of the atmosphere, yet it is the source of all the carbon for plants, and therefore also for animals. Its level is small, but very significant.

The reason for concern is that carbon dioxide gas tends to trap heat radiation (the same "infrared" radiation, invisible to the eye, that is used in heat lamps). This trapping is called the *greenhouse effect* after its similarity to the warming in greenhouses. A more familiar analogy for many people would be the "car-in-the-parking-lot effect": sunlight can get in through closed windows, but hot air can't escape; heat is trapped and the temperature inside rises.

Carbon dioxide isn't the only gas that traps heat. Water vapor is a bigger factor, but so much water from oceans, rivers, and lakes is in contact with the atmosphere that humans exercise no direct control over the water vapor content; the water vapor level is determined primarily by the temperature of the water and air. Methane is also important. Oxygen, nitrogen, and argon, the main constituents of the atmosphere, are virtually transparent to heat radiation, so they play only an indirect role.

The atmosphere has contained carbon dioxide for eons, so the greenhouse effect has been working for eons. In fact, if there were no carbon dioxide or other natural greenhouse gases, physics calculations say the entire Earth would be below freezing. It is the blanket of greenhouse gases that keeps the Earth warm enough to prevent the water in the oceans from becoming ice.

In the last two centuries, humans have added enough carbon dioxide to make our atmospheric blanket a little more effective, and therefore to raise the temperature a little bit. How much? It's very hard to tell because other things also change the climate, particularly solar variability and normal fluctuations in the ocean currents such as El Niño.

The best record of Earth's temperature for recent centuries comes from thermometer measurements. These date back to 1724, when Daniel Fahrenheit invented the mercury thermometer. Later that century, the first measurements in the United States were officially reported to the federal government by Benjamin Franklin and Thomas Jefferson. World coverage was meager back then, but we can still estimate the average temperature of the entire Earth by applying modern statistical methods to the sparse data.[8]

When I began to write this book, three major scientific groups had analyzed the thermometer data.[9] Their results were summarized by an international commission set up to study climate change, the IPCC (Intergovernmental Panel on Climate Change). You don't have to remember what IPCC stands for, but you do need to know those initials. Every few years the IPCC issues a large, detailed report. In 2007 it was awarded the Nobel Peace Prize (shared with former Vice President Al Gore) for its work on climate change. When people say that there is a "consensus" on global warming, it is the IPCC they're referring to. The IPCC conclusion in its most recent report (2007) was that the global warming of the previous 50 years had been about 0.64 degrees Celsius, and that "most" of this warming was anthropogenic—that is, caused by humans—primarily through the greenhouse effect. In those same five decades the average land temperature had risen 0.9°C. (Land warms more than ocean because land heat stays concentrated near the surface; in the oceans, waves mix the water and dilute the heat to depths of 100 feet and more.) The IPCC also said that the globe had been warming since the late 1800s, but that part or all of that earlier warming could have come from changes in the sun's intensity; the anthropogenic contribution for that period could not be determined.

Is the IPCC value for global warming of 0.64°C over 50 years less than you thought? It is barely more than one hundredth of a degree per year. It is indeed such a small temperature rise that it is difficult to detect except by scientists analyzing lots of data. Many people are surprised to learn that the anthropogenic warming is so little so far, even according to the consensus. I polled my class recently and no one knew (or guessed) that it was so small. But it isn't the amount of the warming so far that concerns most scientists (including me); it is the threat of much greater warming in the future.

Although the IPCC analysis seemed plausible, its implications were huge, and we needed to be certain that their estimates were correct. Overreaction to the warming threat could harm energy security and perhaps even the US economy. So it is essential to be sure, to know the magnitude of the danger accurately. Some thoughtful skeptics put forth serious criticisms of the IPCC conclusions and argued that the IPCC had overestimated the warming. They said the IPCC had ignored the dangers of data selection bias and poor temperature station quality, and that they had likely misjudged the increase. Although other measurements, such as sea level rise, confirmed that some warming was taking place, they were indirect and could not distinguish between strong and moderate rates. A proportionate response to warming depends critically on knowing the rapidity of the rise.

In 2009, my daughter Elizabeth and I (we were already working together on a high-tech consulting company we had created) decided to organize a new scientific study to address the uncertainties. We eventually named it the Berkeley Earth Surface Temperature Project. (Some people referred to it as the BEST Project, but we avoided that acronym.) More information can be found at www

.BerkeleyEarth.org. It was to be done in a completely open and transparent way; our analysis would be published and available to all. We organized it under the auspices of Novim, a nonprofit think tank located in Santa Barbara, California. We recruited top scientists, including Saul Perlmutter (who was soon to win the Nobel Prize in Physics for his prior work in cosmology), Art Rosenfeld (renowned for his international work on energy efficiency and conservation), and Robert Rohde, a young physicist with an unmatched (in my experience) talent for data analysis.

The challenge we set for ourselves was to try to use as much of the available digitized data as possible from over 40,000 recording stations around the world. The prior teams had used fewer than 20% of the available stations, primarily ones that had long continuous records. We feared that such stations might be affected by changing environment: thermometers that were once rural could show excess heating over the decades as cities expanded to surround them. We had to handle 1.6 billion temperature measurements, merge 14 different data sets in nearly that many different formats, and develop new statistical methods; most of this was accomplished by remarkable innovations made by Rohde. Moreover, we were determined to put the data in a format that could readily be used by other scientists, so that many more people could do independent analysis. We have now accomplished that; the data can be downloaded at www .BerkeleyEarth.org.

The media labeled us as a bunch of "skeptics." Why reanalyze data when the previous work had already been certified by a Nobel Prize? I even wondered myself—was I a skeptic? I didn't think so, at least not in the sense the critics meant. I thought scientists had a *duty* to be "properly skeptical"—a term that our statistics advisor David Brillinger liked to use. On the other hand, skepticism must be balanced—a scientist should never be so skeptical as to be inconvincible. And skepticism carries with it a danger: that it can be taken as a call to inaction—let's wait until the scientists all reach 100% agreement, until there is no doubt left. Yet without proper skepticism, knowledge will not progress.

Some stigmatized all who questioned the IPCC as "deniers," people who ignored clear and incontrovertible evidence. They suggested that global warming skepticism was a campaign against science itself, analogous to the denial of the theory of evolution. I felt such derogations were improper, and that many who attacked the skeptics were exercising the very antiobjectivism that they claimed to be opposing. I was upset that some groups, including the American Physical Society (in which I had been elected to "fellowship"), solicited signatures on petitions affirming the validity of climate science. Affirm climate science? Does doing so mean that the criticisms of the skeptics are unworthy of consideration? I discovered that several physics professors at Berkeley had signed this petition, and I asked them why they had chosen to dismiss station quality bias and why they thought station selection bias was no problem. In each case, they told me that they had hadn't heard about these issues, and they certainly couldn't dismiss them. They had signed the petitions because they thought their support was needed to defend science, not because they had actually looked at the complaints. Despite the impression conveyed by their signatures, they had not used their scientific expertise to reach the conclusions they were affirming.

The major complaints about the prior work that I considered valid were as follows:

1. Many of the thermometers were poorly placed, some near buildings, some near asphalt pavement, some near heat sources. Such thermometers could show heating even in the absence of true global warming.
2. The analysis groups had "adjusted" the raw data prior to analysis. They had done this to account for changes in instruments, locations, and methods of recording. The net effect of the adjustments was known to increase the estimate of the temperature rise. Had bias slipped in? If done incorrectly, such "corrections" could indicate more heating than had actually taken place.
3. Cities have become hotter as a result of non–greenhouse effects, such as extra energy use and sunlight-absorbing surfaces such as

asphalt. This additional warming is called the "urban heat island effect." Since a large fraction of the temperature stations are located in cities, this heating may have increased the global warming averages by more than its proportionate share.

4. Not all the data were used, many stations were ignored, and the selection was made in a biased way (for example, by choosing stations that had long records) that could have caused an overestimate of the amount of heating.

The Berkeley Earth team worked hard to address these issues. We avoided data selection bias by including all the usable data—which turned out to be 36,866 stations—over five times as many as the 7,280 stations in the "global historical climate network" list used by the prior groups. Doing that wasn't easy; it required the application of sophisticated statistical methods that could combine short records. We tested station quality bias by analyzing the temperature for the good stations and bad stations independently. We tested urban heat bias by analyzing only those regions that were "very rural"—that is, far from any urban location—and comparing the answer with the one that included all land regions. We avoided data correction bias by not doing any corrections; if there was a suspicious change, we tagged that location and sliced the data into two records at that point.

Finally, in late 2011, we were able to use the data to deduce a detailed land surface temperature record, one that was not biased by the effects listed above. Our unexpected (to me) result: we confirmed the IPCC temperature rise. Thanks to our methods, we were able to obtain enhanced accuracy and reduced uncertainties, but the answer was very close to that obtained by the previous groups, about 0.9°C over land. In the end, we concluded that none of the legitimate concerns of the skeptics had improperly biased the prior results. That conclusion suggests to me that those groups had been vigilant in their analysis and treated the potential biases with appropriate care.

But we did get some completely new and very exciting results.

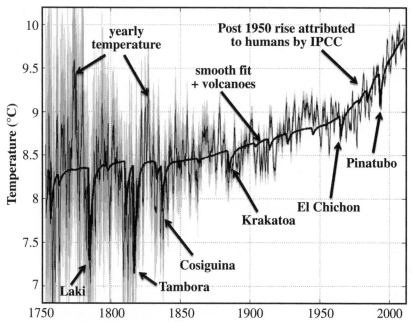

Figure I.6. Average global temperature rise over land from 1800 to the present. The thin line shows the temperature estimates; the gray regions show the 1- and 2-standard deviation uncertainty limits. The dark line is described in the text; it is a smooth curve with dips derived from measurements of large volcanic eruptions.

Our statistical methods, developed largely by Rohde with guidance from Brillinger, enabled us to extend the record nearly twice as far back in time as had previously been done, to 1753, and that long record allowed us to draw some remarkably strong conclusions. Our plot of the yearly land surface temperatures is shown in Figure I.6.

For now, ignore the thick line labeled "smooth fit" that runs across the middle. The thin line that bounces rapidly up and down represents our calculated yearly average of the Earth's land temperature. The light gray area surrounding it shows the yearly uncertainty—the fact that limited statistics or limited geographic coverage meant we could not determine the temperature precisely. The uncertainty limits are tiny in the modern era (so small that they are not visible in the plot), when there is extensive coverage and thousands of thermometers, and large for the early temperatures, when there was sparse coverage and fewer than 100 stations. But even with these large uncertainties, in the late 1700s and early 1800s we can see large significant dips in the temperature and swings more rapid than any that appeared in the modern era. We recognized these as short-lived cool periods from volcanic eruptions.[10]

The sudden dips that happened around 1783 and 1815 are identified with the huge eruptions of Laki and Tambora. Explosive volcanism can spew millions of tons of sulfate particles into the stratosphere, where they linger for several years reflecting sunlight and cooling the ground. The explosion of Tambora was so large that this eruption was blamed for making 1816 into the famous "year without a summer," with deaths around the world from cold spells, crop failures, and famine. There are other dips near 1883 (Krakatoa), 1982 (El Chichon in Mexico), and in 1991 from the explosion of Mount Pinatubo. If you are old enough, you may remember the beautiful "Pinatubo sunsets" that lasted for about two years after that eruption.

When volcanic sulfate particles fall to Earth, some of them get trapped in glaciers forming in Greenland and Antarctica. Those measured deposits are shown by the dips in the solid heavy line of Figure I.6. The close match between this ice core sulfate and our observed temperature drops confirms that despite the small number of stations, our statistical methods do indeed allow us to determine meaningful temperatures as far back as the late 1700s.

Now let me explain the smooth part of the heavy line, the one that runs through the middle of most of the data. Initially, when the Berkeley Earth team saw the relatively steady rise in the temperature from 1753 onward (ignoring the volcanic dips, which are temporary), I tried fitting the data to a smooth curve. First I used a simple parabola and got a pretty good fit. Later I tried other functions, including an exponential and a fifth-degree polynomial. Rohde, however, went in a different direction. He has a great knowledge of global warming data; in fact, he was the creator of a famous project called Global Warming Art that tracked down data and plotted it, putting the results on the Internet, particularly Wikipedia, where his graphs became the best source for numerous scientists looking for things to show in their talks. Rohde redid my analysis but instead of using an abstract mathematical function, he used the known and measured carbon dioxide data curve for his fit. Of course, carbon dioxide is measured in parts per million, not in temperature, so he allowed a variable scale to optimize the match. To my astonishment, the resulting two-parameter fit was closer to the data than were any of the functions I had tried. In fact, Rohde's CO_2 fit, with the volcanoes added, is what I show in Figure I.6 with the heavy line marked "smooth fit." (There is a logarithmic adjustment I'll describe in a moment.)

I have analyzed a great deal of data in my career—from particle physics (in which I earned my PhD) to astrophysics and cosmology to geophysics (in which I wrote a technical book on the origins of the ice ages). But it is rare that in science you find agreement as close as had been found by Rohde. The agreement is astonishing.

Rohde also tried adding the solar variability, measured by sunspot numbers, to see how much it would improve the fit. It didn't, at least not in a statistically significant way. The optimum-fit combination of carbon dioxide and sunspots set the sunspot component statistically equal to zero. That too was surprising to me. It seemed to be telling us that the warming was entirely due to carbon dioxide, and that the solar variability, which previously had been used by the IPCC to account for much of the rise prior to 1950, was unimportant.

In retrospect, the absence of a sunspot component makes sense. Satellite measurements of solar power show that it doesn't vary much even as sunspot cycles come and go. Sunspots are indeed dark, but when they appear, their energy seems to be diverted to the bright parts of the sun. Large numbers of sunspots do not lead to a significantly cooler sun. Our fit shows that one could ignore these cycles and get an excellent explanation of most of the data considering only carbon dioxide and volcanoes.

Why had nobody previously noticed the close agreement between the CO_2 curve and the observed warming? The reason is mathematical; they didn't have our 250-year coverage and our reduced uncertainties. Only by going back further in time with usefully small error estimates do we have enough of a time span to distinguish a fit to CO_2 from the alternatives. Only with a long and accurate record can we test—and rule out—a major contribution from solar activity.

Of course, there are still unexplained swings up and down around that curve, the oscillations that last only a few years. Some of these can be attributed to El Niño, the famous fluctuations in the ocean temperature of the equatorial Pacific. However, in our papers we showed that these are more closely related to the behavior of the North Atlantic. We found a excellent match between these rapid variations and the temperature estimates of that ocean. This match suggests that it is fluctuations in the flow of the Gulf Stream that could be the most important driver of these variations. That is still a hypothesis, not yet a scientific conclusion.

The exquisite agreement between the warming and CO_2 suggests that most—maybe all—of the warming of the past 250 years was caused by humans. That's a remarkable conclusion for someone (me) who had been stigmatized as a skeptic by the media. Yet my opinion had never *changed*. I initially didn't have an opinion because of the many unanswered questions. I didn't know the answer prior to the hard work of our Berkeley Earth team. I had always wanted to keep an open mind and to be persuaded by data and objective analysis, not by consensus pressure. My opinion didn't change; it developed.

As our project began to show the reality of global warming and its human origins, newspapers printed ridiculous articles about how we had reached conclusions opposite to those that our supporters had wanted. We had received financial support from a wide range of foundations, including those created by Lee Folger, Gordon Getty, Bill Gates, Bill Bowes, and Charles Koch. But these foundations had all made it clear to Elizabeth and me that they were not "hoping" for any particular result—and indeed, if they had been, we might not have accepted the funding. To the contrary, they told us that they were supporting Berkeley Earth because they felt that we were open minded and objective, that we would do a good job and help to resolve important questions about climate change.

I'll now use the Berkeley Earth results to try to predict future warming. According to a straightforward physics calculation, the additional greenhouse effect of the CO_2 is not proportional to the CO_2 level but to the logarithm of the CO_2 level. (The physics reason has to do with the fact that most of the effect comes from the edges of the CO_2 absorption lines; these broaden only logarithmically. This response is verified by the complicated climate models, and it is not seriously disputed.) In fact, because of this behavior, Rohde's function in Figure I.6 gets its smooth part not from the concentration of CO_2 but from the logarithm of that concentration.

For predictions of greenhouse warming, it is conventional to compare climate models by asking what they predict for the time when the CO_2 reaches twice the preindustrial levels—that is, twice 280 parts per million. If we assume continued exponential growth in CO_2, that should happen about 2052. A logarithm calculation[11] then predicts for that year a land temperature 1.6°C warmer than now and a global temperature (land + oceans) 1.1°C warmer than today. What will happen afterward? Here's an insight for the mathematically inclined: If CO_2 grows exponentially, and the greenhouse effect increases logarithmically, then the warming should grow linearly. (That's because if you take the exponential of a number, and then the logarithm of that, you get back the original number.)[12] So double the time interval and you'll predict double the temperature rise. This means that about 40 years after 2052, there should be another 1.6°C rise over land and 1.1°C rise worldwide. And so on, every 40 years, until we stop dumping CO_2 and other greenhouse gases into the atmosphere.

These calculations are crude, but their simplicity is their virtue. Some people complain that the sophisticated IPCC models are so complex that they hide assumptions deep inside computer code, that some of these assumptions may be unwarranted or biased, and that the models may have been "adjusted" by programmers seeking a preconceived answer. In contrast, our log calculation is completely transparent, and it gives estimates that are not substantially different from the complex computer-based ones. For the smooth part of the curve we used only two parameters, an offset and a scale. It is actually just *one* parameter if we adopt the usual convention in the field and set the offset equal to zero for both the data and the fit. To get such good agreement with one parameter is remarkable. It is the sort of "back of the envelope" computation that physicists love.

Some might argue that this calculation ignored water vapor feedback, methane, snow cover, aerosols, and clouds. To the extent that their changes roughly follow the CO_2 changes, then our use of the *shape* of the CO_2 curve with adjustable amplitude can be thought of as a proxy for all such effects. To test this, I tried running the program with methane explicitly included; indeed, the results were similar, but the temperature change was shared by the two components. Of course, you can always get a better fit by adding in more parameters. We could add in an aerosol term, and that can only improve the agreement.

Conclusion: It appears that human (anthropogenic) effects can account for all of the trend in warming that we have observed over the past 250 years. Unless some new phenomenon kicks in (clouds increase suddenly; the Chinese economy stalls; we hit a "tipping point"), this model suggests that we are likely to see additional 1.1°C (world) and 1.6°C (land) temperature rises by 2050.

Tipping Points

If it were strong enough, positive feedback could lead to instability and runaway warming or other rapid and disastrous changes. Runaway greenhouse warming is responsible for the hot surface conditions of Venus—about 900°F—on a planet that is otherwise very similar to the Earth. Mechanisms for such catastrophes are called *tipping points*, and several possible tipping points for the Earth have been studied:

- The Antarctic ice sheet loosens and slips into the sea (triggering over 100 feet of sea level rise around the world).
- Freshwater from Greenland melting disrupts the Gulf Stream and changes sea current flow all the way to the Pacific.
- Methane, a potent greenhouse gas, is released from melting permafrost leading to runaway warming.
- Methane is released from the seabed because of warming Arctic waters.

None of these are currently considered imminent dangers. Methane from permafrost frightened a lot of people, but it turns out that the methane that is seeping up comes not from the permafrost itself but from sources deep below that would not produce more if the surface permafrost melted. The Greenland melting scenario caused some deep concern until the original proponent of the theory, Professor Wallace Broecker of Columbia University, showed in calculations that it was not a serious danger after all. Perhaps the tipping point that is most worrisome is the warming of the Arctic waters, but that would take a larger increase in temperature than we expect in the next century.

In my mind, the scary thing about the tipping points is not that any are considered a current danger, but rather that smart people keep thinking up ones that we hadn't previously recognized. We do not understand climate well enough to be sure that we've identified all of the major contributors. The really dangerous tipping point is the one that is real and has not yet been recognized.

Negative feedback is also possible, and that could *reduce* the expected greenhouse warming. Suppose, for example, that the increased water vapor from warming creates more clouds that reflect sunlight. A mere 2% increase in cloud cover is enough to negate the expected warming from a doubling of carbon dioxide. (Recall that so far, the level of carbon dioxide has gone up 40%.) According to the IPCC summary, cloud cover has not yet changed. However, poor understanding of cloud behavior is the biggest cause of uncertainty in the warming predictions.

Global warming could trigger other changes in climate, including changes in the number and intensity of storms, a rise in sea level, shifts in the distribution of rains, and other effects in fragile areas such as rain forests and coral reefs. Some fertile areas might go dry; some dry areas might get new rainfall. Not all effects are bad; the carbon dioxide could enhance plant growth (it is frequently injected into greenhouses to do just that), and some people living in Canada have expressed their desire for a slightly warmer climate. But most people believe that, like genetic mutations in evolution, most abrupt changes in our climate would be bad for most people.

Local Variability

Despite claims that climate change is "clear and incontrovertible," the evidence is actually quite subtle. It takes careful scientific analysis to see it. As an individual, you really can't sense it yourself. If you think you can, you are likely confusing global climate with local weather. Take a look at the map of weather stations in Figure I.7. It shows all of the stations in the United States and nearby that have been recording for at least 100 years. Stations whose temperature has gone up are marked with plus symbols; stations with temperatures that have gone down are marked with circles. Notice that about a third of the stations actually recorded a *decrease* in temperature over their lifetimes.

This seems crazy. How can we have so many stations cooling when the world is warming? But if you count, you'll find that although one-third of the stations are cooling, two-thirds are warming. The same fractions hold when we look at the 36,866 stations located on all seven continents around the world. Global warming is not evident in the individual stations, but it is in the average.

The reason a cooling trend is registered at some temperature stations, even though globally the climate is warming, is that local climate is quite variable compared to the average global climate. If

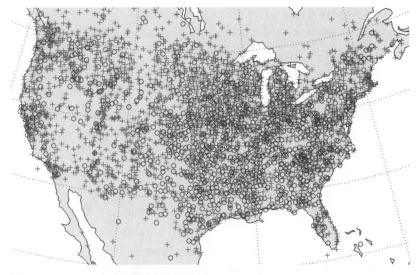

Figure I.7. Map showing temperature stations in the United States and neighboring countries that have been recording for at least 100 years. Locations with rising temperatures are marked with plus signs; those with cooling temperatures are marked with circles. One-third of the stations have cooled; two-thirds, warmed.

and when human-caused global warming reaches 2°C or 3°C, the local effects will not be sufficient to cancel it, and virtually all of the world's stations will indicate warming.

I sometimes suspect that this variability and the large number of locations that have cooled encourage many climate change skeptics. One-third of all people live in areas where the temperature has dropped over the past century. Global warming is certainly not evident to them. But people who live where the temperature has risen may also be fooling themselves. I've heard a congressman state, "Global warming is so obvious that all you have to do is step outside and feel it yourself." Yet temperature stations show that parts of this city had actually *cooled* over the past 50 years. Perhaps when the congressman spoke, he was referring to his hometown; either that or he was overreacting to a hot day (and indeed it was hot in Washington that day). In fact, it is not possible to detect the current small world-average 0.64°C rise without hundreds or thousands of records.

You may have been surprised when I said that global warming has not been noticeable except to scientists. What about the other aspects of climate change, such as the increase in hurricanes? In tornadoes? In wildfires? What about Hurricane Katrina? What about the melting of Alaska?

None of those phenomena are evidence for human-caused climate change. Linking them to climate change is not scientifically sound, although it is often done by people (including scientists) who are so genuinely concerned about the dangers of the greenhouse effect that they exaggerate either inadvertently or sometimes on purpose. The problem with such exaggerations is that they provide ready targets for critics of the global-warming theory. In my opinion, the temperature data are solid. Most of the other effects are all either wrong or distorted. So let's review some of the key claims.

Hurricanes

Figure I.8 shows the number of hurricanes hitting the United States every decade since 1850. The chart seems to contradict the common belief that the number of hurricanes has been increasing. In 2005, for the first time ever, we actually ran out of letters of the alphabet and had to recycle letters to have enough for the names in that year's hurricane season. But the reason we used up the alphabet is that we now detect hurricanes far out at sea that are in regions where ships rarely travel but are accessible by satellite photos and buoy wind measurements. The fact that we're *observing* more storms doesn't necessarily mean that more are occurring.

To compensate for such observational effects, scientists like to pick an "unbiased subsample." Storms that impact the US coast are ideal because they have been reliably reported since the early 1800s. When we use such a subsample, as we did in Figure I.8, we see that

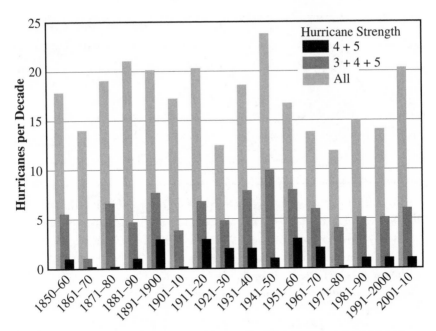

Figure I.8. Number of hurricanes hitting the United States every decade. The light bars show all hurricanes; the small black bars in the fore-ground indicate the most intense hurricanes (categories 4 and 5), and the intermediate hurricanes are shown by the bars in between. The rate of hurricanes hitting the United States has not been increasing.

there is no increase. Look at the maps in Figure I.9, which show the Atlantic hurricanes in 1933 and in 2005. These maps clearly reveal the danger of data selection bias. The number of storms seems to be increasing, but the truth is simply that we have better and more widely distributed instruments, so we are detecting storms that we previously would have missed.

What about Hurricane Katrina? It is widely thought of as a cat-egory 5 storm (the most intense kind), but it was actually only a category 3 storm by the time it hit New Orleans. That city is a very small target that had not suffered a direct hit in recent decades; the first moderate hurricane that struck—Katrina—did little immediate damage, but much of New Orleans is below sea level and the next day the levees failed. The destruction of New Orleans cannot be

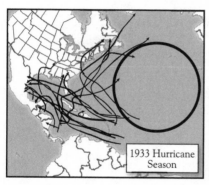 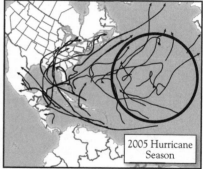

Figure I.9. Hurricane tracking maps based on those of Chris Landsea of the National Hurricane Center,[13] showing two of the busiest hurricane years on record for the Atlantic: 1933 and 2005. The circles highlight open-ocean areas showing large differences in activity. Note that in 1933 no hurricanes were detected in the deep Atlantic. However, few observations were made in that region, so we can't know whether hurricanes were there or not.

attributed in any meaningful way to global warming—even though temperature records show that global warming is real.

Tornadoes

The year 2011 was marked by a spate of particularly destructive tornadoes, killing 482 people—117 in Joplin, Missouri, alone. But that record death toll doesn't reflect a huge rise in the number of tornadoes; rather, it reflects the bad luck that in 2011 the tornadoes happened to hit population centers. Figure I.10 charts the number of intense tornadoes (categories F3–F5) in every year since 1950. If you look at the trend, you'll see that the rate of tornadoes has been going down; statistical analysis verifies this conclusion.

Does the gradual decrease in intense tornadoes in the United States contradict global-warming theory? No; the theory never predicted that violent storms would increase; it predicted only that they *might* increase. The argument for increase is that warming means that more energy is available to create storms. The argument for decrease is that in most theories the poles warm more than the equator, and that means that the temperature difference from north to south gets smaller; in many theories it is this temperature *gradient* that generates storms, not the absolute temperature.

A look back at the temperature station map (Figure I.7) reveals that in most southeastern US locations the temperature hasn't risen at all in the last 100 years. Yet this is the area in which people attribute tornadoes to warming.

Polar Warming

What about Alaska and Antarctica? Surely the melting of these polar regions is strong indication of global warming, right?

The evidence is not as compelling as most people think. In 2001, the IPCC recognized that the upcoming GRACE satellites would be

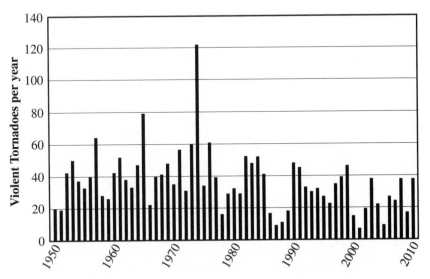

Figure I.10. The number of intense tornadoes from 1950 to the present.

able to make remarkably precise estimates of ice in Antarctica. An acronym for "Gravity Recovery and Climate Experiment," GRACE is an exquisite system that measures small changes in gravity—such as that from increased or decreased ice. So in 2001 the IPCC asked the climate modelers to estimate the expected yearly changes in Antarctic ice. To some people's surprise, all the models predicted that global warming would cause the ice to *increase*, not decrease. In retrospect, the reason was obvious. Even with warming, Antarctica stays below freezing, but warmer oceans have more evaporation, and that leads to more snow.

When the satellites flew and the measurements came in, they contradicted the models. Antarctic ice was decreasing, at the rate of 36 cubic miles of ice every year! How could the models get it so wrong? The answer, unfortunately, is: easily. Even though it looks big on Mercator projections, Antarctica covers only 2.7% of the globe, and the programs are not able to make reliable predictions for such tiny regions. Microclimate is much harder to predict than global climate.

So the modelers looked at their models, made changes, and now they reproduce the decreasing ice of Antarctica. Is that a success of global-warming theory? No; it just shows the limitations of modeling programs, and the "tuning" that is continually necessary to match existing records. But it does mean that the melting of Antarctica *cannot* be taken as evidence for global warming. Predicting effects after they are seen is easy even for wrong theories; we must judge models by what they predict *before* the answer is known.

The models did predict that Arctic ice would decrease, and that effect has been observed, as the computer images of Arctic ice in Figure I.11 indicate. Does that prove the models are right? Unfortunately, the models are qualitative, not quantitative, so it is hard to verify them in any detail. And we have to be very careful here, lest we fall into the trap of "cherry-picking." (At a grocery store, we pick out all the best cherries and buy them; someone seeing our selection thinks all the cherries this week are great.) Recall, the

Figure I.11. These composite satellite images of the Arctic region show the reduction in ice coverage between 1979 and 2003.

prediction for Antarctica was wrong. If we judge a model only on its successes and ignore the failures, then every model will be deemed a success.

There are other effects that could be responsible for the melting of Alaska and the Arctic. Decadal oscillations in the north have been observed in old measurements of sea surface temperature and pressure. They have the following names: "Pacific Decadal Oscillation," "North Atlantic Oscillation," "Atlantic Multidecadal Oscillation." You can read all about them in excellent Wikipedia articles. These oscillations are not completely understood, but they could be ocean instabilities similar to the one in the equatorial Pacific that gives rise to the El Niño cycle. Is it possible that this is what is happening in the north? It's hard to rule out.

Some evidence does suggest that the Arctic ice goes through cycles too. Roald Amundsen and a crew of six navigated through

Figure I.12. The Northwest Passage was clear of ice in 1903–06 when Roald Amundsen and his crew took this wooden boat, the Gjøa, through the Arctic Ocean north of Canada.

the Northwest Passage waters north of Canada in the years 1903–06 using a small boat (47 tons, sail plus 13-horsepower motor) named Gjøa (pictured in Figure I.12), hardly an icebreaker.

The Hockey Stick

One chart that my colleagues at Berkeley had found particularly compelling was the "hockey stick" graph. The hockey stick graph is dramatic. It shows that current climate is truly unprecedented, at least for the past millennium, with a dramatic rise starting around 1970. A famous version of this graph is shown in Figure I.13. It appeared on the cover of the World Meteorological Organization's 1999 report titled *On the Status of Global Climate*. This plot shows estimates of the temperature of the Earth going back 1,000 years. Of course, thermometers weren't accurate until the 1700s, and there was not good worldwide coverage until the twentieth century, so the estimate for the earlier centuries is based on indirect indicators of temperature, including tree ring widths, coral growth, and other indirect "proxies."

Forget all the other evidence; the recent rise shown in this plot is clear and incontrovertible. The present warming is unlike anything that happened in the previous thousand years. Well, not quite. The scientists who published this plot were not being completely candid. The original data, plotted in Figure I.14, didn't show such a dramatic recent rise.

The scientists responsible for the published graph felt that showing the recent proxy data, some of which dips, might mislead the reader into thinking that global warming was not as bad as they thought it was; we know this because their e-mails were leaked in what came to be known as the "Climategate" scandal. One of the proxy sets used to estimate past temperatures seemed to show a decline in temperature in the last 40 years, and the scientists knew this decline was not true; temperature measurements from thermometers show a clear rise in this period. Why the disagreement? My best guess is this: proxy data are just not very reliable. If you look at the period from 1600 to 1700, you'll see that the three different curves, from three different projects, disagree about as much as they do in the last 40 years. Tree ring growth responds to things other than temperature, including rainfall, humidity, cloud cover, and possibly wind.

What does a scientist do with such discordant data? I earned my degree in particle physics and later became well known for work in astrophysics, and in these fields the tradition is clear: you show the troublesome data, you explain why you think the data are unreliable, and then you describe what happens when you ignore or "correct" those data. You must work hard to convince your intended audience—skeptical scientists—that you're doing all this in an unbiased way.

What is improper is to *hide* discordant data. Unfortunately, that's exactly what the group publishing this temperature plot chose to do. They cut out the proxy data from the last 40 years and replaced it with thermometer data. (I emphasize that, according to my scientific training, replacing proxy data is OK—as long as you show

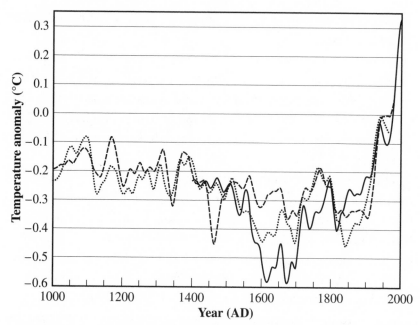

Figure I.13. *Estimated temperature trends over the last 1,000 years, as published by the World Meteorological Organization in 1999. Most of the plot was based on "proxy" data—indirect estimates such as from tree ring widths and coral growth rates; the recent data (post-1960) came from thermometers. The three plots show the analysis of three groups analyzing different selections of the data.*

the removed data so that other scientists can judge whether you did the right thing.) Deep in their papers they said what they had done, but they did not show the cut data, and they refused to release it when asked to do so by skeptical scientists. The published plot became famous, and few people who saw it realized how it had been "adjusted." Although the plot was well known to scientists, the adjustments were not.

Sea Level Rise

After all these invalid indicators of climate change, let's look at one that really is correct: the rise in global sea level. According to the

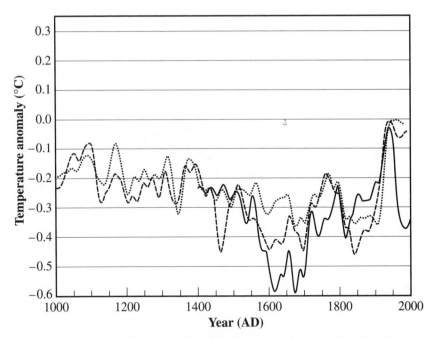

Figure I.14. This is the same plot as in Figure I.13, but it includes the proxy data from 1960 onward that had not been included in the originally published plot. This plot does not include the thermometer data that had been used to replace the recent proxy data.

IPCC, the level has risen 8 inches in the last century. Sea level rise is well documented, as shown in Figure I.15. Most of that rise comes from the simple fact that water expands when it warms. In addition, some of the rise can be attributed to melting glaciers around the world.

The problem with interpreting sea level rise is that it is difficult to know how much is due to natural causes, how much is due to local pollution; for example, much of the melting of Greenland is believed to be caused by soot—something that can be controlled much more easily than can carbon dioxide. Sea temperature increases more gradually than air temperature, simply because there is so much more mass in water; just the top 32 feet of the oceans weighs as much as the entire atmosphere. It is very hard to know how big the lag is, so it is difficult to use these data to verify or to throw doubt on the warming calculations.

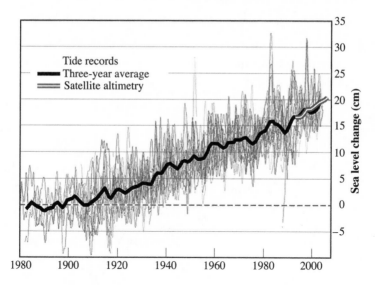

Figure I.15. Records of tide levels show a clear rise in sea level in the last 130 years.

Another issue is the absence of a clear change in the slope of sea level rise. The plot of land-surface temperature (Figure I.6) shows a rather abrupt and sudden rise starting in 1970. The rapid increase in temperature is a good match to the increase in atmospheric carbon dioxide from fossil fuels (the dark line). But no such slope change is evident in the sea level plot. What we really want to know is the human component of global warming, and the lack of such a slope change means it is difficult, maybe impossible, to use the sea level data to confirm that most of global warming is caused by humans.

It is worth pointing out that the sea level rise so far has been only 8 inches (20 centimeters). Of that, the US contribution is estimated to be about a fifth, meaning that we have contributed less than 2 inches of the total rise. To go one step further, our automobiles have contributed about a quarter of that—less than ½ inch. This number will be useful when you consider spending money on new generations of automobiles.

According to the IPCC, through the remainder of this century the level of the sea is expected to rise another 1–2 feet. That's enough to affect the coast of Bangladesh and some islands in the Pacific, but it will not flood Florida or New York, as imaginatively depicted in the movie *An Inconvenient Truth*. The kind of flooding shown in that movie depends on an immediate melting of most of Greenland or Antarctica—events that the IPCC does not expect to happen within the next 100 years.

Assuming It Is a Threat—Can We Stop Global Warming?[14]

Can we stop global warming? The answer—which is quite surprising—is one that a future president needs to understand. For the sake of this discussion, let's assume that the theory of global warming is absolutely correct. If you don't accept that, just bear with me. I'll begin with a little background.

Every few years there is an international meeting on climate change. Representatives of most of the countries of the world come together to discuss the issues, and to try to reach an agreement on what to do. In 1997, then US Vice President Al Gore attended the meeting held in Kyoto, Japan, and he signed the treaty on behalf of the United States. Gore agreed that the United States would cut greenhouse gas emissions until they were 5% lower than the 1990 level. Signing the treaty didn't commit the United States; it still had to be ratified by the US Senate. It never was.

In 2009, President Barack Obama attended the climate change conference in Copenhagen, Denmark. There had been many preparatory meetings, and a proposed treaty was on the table. Many people expected President Obama to sign it. He didn't. Some people blamed Obama's political opponents, the Republicans, for not having promised support. But treaties are signed before they are ratified; an agreement is reached, and then the president tries to win over Congress. In fact, President Obama's stated reason for rejecting the treaty was that China had refused to agree to inspections.

To understand why inspections were so important and what was going on, let's consider the treaty that was under consideration. According to this agreement, the United States would cut greenhouse emissions by 80% by the year 2050; the rest of the developed world would promise to cut their emissions by 60%–80%; and China and the developing world would pledge to reduce greenhouse emission intensity 4% per year, resulting in a 70% intensity cut by 2040. (Please notice the word *intensity* in the last sentence.)

Now let's be really optimistic. Assume that President Obama had won over the Republicans and the treaty was ratified. Assume that *all* parties—the entire world—abided by it. Assume that China allowed inspections, and we were satisfied. Assume all the countries of the world achieved these stupendous goals! Atmospheric carbon dioxide emissions would drop dramatically, right?

Wrong. Very badly wrong. Under the treaty, carbon dioxide emissions would continue to rise to nearly 4 times the current level. Total atmospheric carbon dioxide would rise above 1,000 parts per million by 2080 (the present fraction is just above 390 parts per million), and—if IPCC climate models are right—global temperature could increase by more than 3°C (5.4°F).

How can that be? The answer is in the numbers and in the technical meaning of the word *intensity*. First the numbers: According to the Netherlands Environmental Assessment Agency, China's CO_2 emissions grew at a rate of 11% per year from 2002 through 2010, and surpassed those of the United States in 2006.[15] The agency's estimates are shown in Figure I.16.

Note that in 2010, China emitted 70% more carbon dioxide into the atmosphere than did the United States. Since its emissions were growing at over 10% per year, while the US emissions were steady or falling, it is likely that by the time you read this, China's yearly greenhouse gas emissions will be double those of the United States, perhaps higher.

Now to the meaning of *intensity*. This is a legal term that means CO_2 emissions divided by the GDP (gross domestic product). The Chinese economy has been growing at an average annual rate of 10% for the last 20 years. If this economic growth continues, then the proposed treaty would require China to reduce its CO_2 growth to 4% less than this. That means that the increase in China's CO_2 emissions would be limited to 6% per year. The future CO_2 emis-

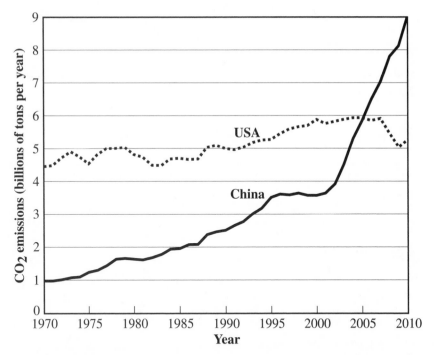

Figure I.16. The production of CO_2 by China and by the United States.[16]

sions, assuming the proposed limits, are shown Figure 1.17. Note how the curve for emerging economies shoots up. Compounded annually from 2012 to 2040, a 6% annual CO_2 growth yields an increase of 511% by 2040. The implications for the Copenhagen Accord are enormous. Most future CO_2 will come from the emerging economies, with China contributing 44%.

China claims already to be cutting yearly CO_2 intensity by 4% as part of its 5-year plan. If those claims were accurate, China's emissions would currently be growing by 6% per year, disagreeing with the Dutch estimates. Chinese President Hu offered at Copenhagen to "continue" the Chinese "reductions." But with the great disparity in estimates, it is essential to have inspections; trust but verify. President Obama insisted on inspections, and President Hu refused.

China is a poor country, and even if we assume its emissions are now double those of the United States, its emissions per capita are only 45% of ours. Isn't it unfair to ask them to cut? Perhaps, but

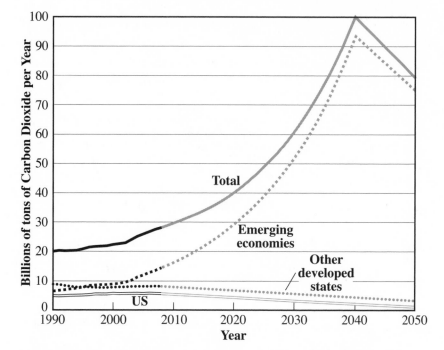

Figure I.17. Greenhouse gas emissions (primarily carbon dioxide and methane), assuming that the proposed Copenhagen treaty had been approved and abided to, and that the developing world continues its high economic growth rate.

warming doesn't come from emissions per capita; it comes from total emissions. China's emissions intensity is now 5 times that of the United States, and even at only a 6% rate of increase (compared to the US rate), its per capita emissions are growing so rapidly that they will surpass US per capita emissions by 2025.

Won't China recognize that global warming is a looming disaster, and cut emissions? Maybe, and in Figure I.17 I assumed they did start cutting in 2040. But energy use correlates with wealth, and wealth is a wonderful thing. China is fighting poverty, malnutrition, hunger, poor health, inadequate education, and limited opportunity. If you were the president of China, would you endanger progress to avoid a few degrees of temperature change? Another complicating factor is that slowed growth in China could trigger political instability.

The Chinese population is migrating in droves from the countryside to the cities, and the Chinese government is expected to provide jobs. My political science colleagues say that the Chinese leaders would feel threatened if the country's growth were to drop below 7% per year.

It hasn't helped that the 2007 IPCC report included a huge mistake. It stated that by 2025 the Himalaya glaciers might melt from global warming, with dire consequences for both China and India. There was no science behind this claim; the referees for the report said it should omitted, but the author (who was able to override the referees) left it in because, he later said, the statement would have a big impact. The chairman of the IPCC, Rajendra Pauchari, pointed out that this mistaken statement was just a tiny part of the report. True, but as far as China and India were concerned, it was the most important statement in the entire report, and it was wrong. The falseness of that statement led to a distrust of the IPCC by people who had previously accepted the panel's findings.

Note that reductions in US emissions (look at Figure I.17 again) achieve little unless China reduces rapid growth. Blaming past US excesses may be accurate (or not), but doing so doesn't solve the problem. If China emulates the United States by cutting emissions only *after* economic growth, we will suffer huge increases in carbon dioxide.

As far as global warming is concerned, the developed world is becoming irrelevant. Every 10% cut in US emissions is completely negated by 6 months of China's emission growth. Setting an example won't help if emerging economies wait until they are emerged before reducing carbon dioxide emissions. By 2040, China could be the most economically powerful nation on Earth, by far. The West can argue and cajole, but it won't be able to impose sanctions or more forceful measures. Temperature will be at the mercy of the newly wealthy world.

Expensive efforts to reduce Western emissions may set an example, but it's a worthless example if the developing world can't afford to follow. Perhaps only *cheap green* that provides economic benefit to the developing nations will be adopted by those nations. Improvements in energy productivity (efficiency and conservation) could significantly reduce future emissions, and yet be profitable. A great reduction would take place if China were to switch from coal to natural gas as its source of energy for electricity; natural gas emits only half the carbon dioxide. Low-cost solar, wind, and nuclear power are good bets. All these must be included under any new, clean development mechanism. To be sustainable in the developing world, green technology must be extremely low in cost, subsidized by the developed world, or profitable. If developed countries harm their economies by expensive emissions cuts, a watchful China and India will never follow that example.

Former Vice President Al Gore and the "350" movement that he endorses (to bring the carbon dioxide level down from the current 392 parts per million to 350) say that whatever treaty we sign shouldn't put constraints on developing nations. But a West-only approach doesn't even address the emissions reduction issue, except to the extent that we will someday be able to say, "At least the warming wasn't *our* fault." In fact, to favor dropping carbon dioxide to 350 parts per million, while arguing for *no* restraints on the developing world, is inconsistent, as Figure I.17 shows. It is ironic that a movement named after a number actually ignores the numbers.

Just to stabilize greenhouse emissions, the emerging economies must cut emission intensity by 8%–10% per year. That will be hard—probably impossible. China has been installing a new gigawatt of coal power plants each *week*—over 50 gigawatts per year. For comparison, New York City uses only 10 gigawatts total. Coal is dirt cheap, and it offers stiff competition to alternative technologies. To address the problem, we need to surge ahead, make carbon reduction in China work, and maybe even subsidize China's effort. A dollar spent in China can reduce carbon dioxide much more than a dollar spent in the United States on a high-tech solution that is too expensive for the developing world to adopt. We can certainly help them develop technology to switch from coal to natural gas.

In its economic planning, China projects that its yearly growth will drop from 10% to 4% by 2040. If that happens, and if the Chinese can abide by a 4% cut in emissions intensity, then their emissions will no longer grow after that—but they will still be much higher than those of the United States.

Is the rapid growth in the Chinese GDP a bubble that will soon pop? This kind of economics question is beyond my expertise to answer, but let me suggest some relevant facts. The Chinese GDP per capita is currently 2% that of the United States—16% if you take into account "purchasing power parity" (the fact that local prices of food, clothes, and housing in China are lower than US prices). China's GDP is rapidly catching up to ours—as I mentioned earlier, at the rate of 10% growth per year for the last 20 years—and until the Chinese match us, such growth doesn't appear to be a classic bubble. It might be unwise to depend on future slowing of China's growth as a solution to the potential danger of global warming. When you are president, you'll have to consult with your economic advisors.

What can we do? Could we learn to live with the heat? Maybe global warming is good. Despite claims to the contrary, storms aren't increasing; the rate of hurricanes hitting the US coast has been steady for 150 years, and the number of damaging tornadoes has been declining. Will Happer, a former director of energy research for the Department of Energy, argues that the past increase in carbon dioxide may have helped the agricultural revolution. And chilly Berkeley (where I live) might be nicer with a few degrees of warming.

As a future US president, you must know the numbers. If you offer 80% cuts in US emissions, you must get 8%–10% of annual cuts in China's CO_2 intensity to level its CO_2 growth. Agree to less, and atmospheric CO_2 will soar along with the economies of the emerging states. We must help emerging nations conserve energy and move rapidly from coal to lower-carbon power sources, including natural gas, solar, wind, and nuclear. Recognize that, like the emperor's new clothes, proposals that make the West bear the burden are less realistic than the people tailoring them may have led us to believe.

Geoengineering

Halting carbon dioxide growth is a daunting challenge. Suppose we fail? What can we do about the climate? We broke it; we should be able to fix it. Because the scale will be huge, involving the entire Earth, the goal of countering the consequences while allowing emissions to increase is called *geoengineering*.

Many methods have been proposed. One of the first was to enhance plant growth in the blue ocean by dumping in iron. This proposal fascinated many people who hadn't previously appreciated that life in much of the ocean is limited by the lack of this key nutrient. Other approaches include cloud-seeding methods to enhance clouds; just a 2% increase in cloud cover could reduce the temperature as much as a doubling of carbon dioxide would increase it. Others have proposed deploying orbiting mirrors above the atmosphere to reflect sunlight.

One fascinating geoengineering idea is to squirt a few million tons of sulfate particles into the stratosphere, where they would form aerosols that reflect sunlight. This approach was inspired by the observation that volcanic eruptions that spew dust and aerosols high into the atmosphere can cool the Earth; in 1991, Mount Pinatubo set an example, and back in 1815 the eruption of Mount Tambora led to the year without a summer. A simple physics calculation suggests that just one pound of sulfates injected high into the stratosphere could offset the warming caused by thousands of pounds of carbon dioxide. The aerosols stay up only a few months, so you have to keep injecting them as long as the carbon dioxide in the atmosphere remains high. But the ephemeral nature of the aerosols can be seen as an advantage: if you discover a horrible side effect, you can stop. Besides, such a high-altitude aerosols will give us pretty sunsets.

Or we could foam up the oceans to increase sunlight reflectivity.

This is going to be a very short section because I don't believe that geoengineering will ever be seriously considered. The idea that we can understand the environment well enough to take the risk of changing things on such a large scale rightly frightens many people. Even if we could reduce overall global warming, could we be sure we would not alter local climate, such as the cycle of the monsoons or the stability of the Antarctic ice sheet? Imagine trying to write an environmental impact report for shooting a few million tons of sulfate aerosols high into the atmosphere.

Some engineers argue that we have to do the basic research now, just in case our carbon dioxide increases bring us to a catastrophic tipping point—one that triggers a runaway warming that threatens all civilized life. Others worry that research in geoengineering can give us a false sense of security, making us think that the danger of continued carbon dioxide emission is not as threatening as we would otherwise believe.

There is certainly one version of geoengineering that would work. It addresses both climate change and the additional danger that atmospheric carbon dioxide is gradually acidifying the oceans. The only question is whether we could afford it. It is the obvious one: leave the fossil fuels underground.

The Global-Warming Controversy

Global warming is both a scientific conclusion and a secular religion. It is important to recognize this dichotomy when listening to the "debate"—because the media tend to emphasize the religious aspects. The religion of global warming (both for and against) tends to be fundamentalist and doctrinaire, so it makes for starker disagreement, and that makes for more interesting stories than the scientific ones.

Some people love to argue, particularly when they sense vulnerability in their opponents. Alarmists have tried to scare the public into supporting strong measures to reduce global warming, but they've done so by exaggerating the facts. I suspect that's why another extreme group has arisen—one I call the "deniers." Deniers believe global warming is bunk; they attack the false arguments of the alarmists and point out the mistakes. But they, too, exaggerate, particularly by cherry-picking. One denier posted online a list of 20 sites that show temperature cooling; I could have sent him 12,000 more. With alarmists fighting deniers, what should have been a discussion turns into a debate, and the media love it. As far as the public is concerned, the knowledgeable people have split into these two groups.

In fact, the political spectrum of climate change is much wider. I like the following categories:

- **Alarmists.** They pay little attention to the details of the science. They are "unconvincibles." They say the danger is imminent, so scare tactics are both necessary and appropriate, especially to counter the deniers.
- **Exaggerators.** They know the science but exaggerate for the public good. They feel the public doesn't find the current level of warming, 0.64°C over the past 50 years, threatening, so they have to cherry-pick and distort a little—for a good cause.
- **Warmists.** These people stick to the science. They may not know the answer to every complaint of the skeptics, but they have grown to trust the scientists who work on the issues. They are convinced the danger is serious and imminent.
- **Lukewarmists.** They, too, stick to the science. They recognize there is a danger but feel it is uncertain. We should do something, but it can be measured. We have time.
- **Skeptics.** They know the science but are bothered by the exaggerators, and they point to serious flaws in the theory and data analysis. They get annoyed when the warmists ignore their complaints, many of which are valid. This group includes *auditors*, scientists who carefully check the analysis of others.[17]
- **Deniers.** They pay little attention to the details of the science. They are "unconvincibles." They consider the alarmists' proposals danger-

ous threats to our economy, so exaggerations are both necessary and appropriate to counter them.

The evidence shows that global warming is real, and the recent analysis of our team indicates that most of it is due to humans. Although the American public showed great interest in global warming a few years ago, subject people to a bit of economic recession and they tend to forget about it. In his 2011 State of the Union message, President Barack Obama didn't mention the issue of climate change. In 2012, his only mention was this: "The differences in this chamber may be too deep right now to pass a comprehensive plan to fight climate change." Some people attribute the public disinterest to a campaign of misinformation, and they are partially right. But the real cause of the apathy may be the discredited claims of the alarmists—those who attribute every weather change to global warming, who predicted many more Katrinas, and who stated back in 2009 that the Copenhagen treaty was our last chance; failure to sign it would doom us irreversibly to horrific climate change. Another reason might be the enormous growth of the Chinese emissions, which makes US unilateral action seem inadequate.

The US public resents being fooled. There is backlash—too much. And now more than ever, they do not think global warming or climate change is high on the list of threats to the United States. In a 2011 Gallup poll, 41% of respondents said that the seriousness of global warming is "exaggerated." In a Reuters poll taken in August 2011, only 37% of Democrats said they believe that global warming is the result of primarily human action, and only 14% of Republicans believed that.[18]

Assuming the theory is right, then the expected warming is very slight—only a few degrees Celsius over the next 50 years—but humans are supersensitive. Just a few degrees can change fertile regions into deserts or melt ice caps. Is global warming an imminent danger for humans? That depends on what you mean by *imminent*. Has it already caused great harm around the world? No, not really. The fear of global warming is primarily concerned with future effects.

Will it eventually cause great harm? That is uncertain. There is a plausible case to be made that future warming will change our climate in a way that is unprecedented for human civilization, and that such disruption is very likely bad. But for a public that is concerned about its own economic well-being, that is an abstruse argument.

Global warming does pose a real threat—one that we need to take seriously, even if it is hard to quantify. The danger is not that human-caused warming is real or that it has already done damage. The concern is that the levels of carbon dioxide in the atmosphere have risen significantly in the last few hundred years, thanks to human activity, up from 280 parts per million to more than 392. That's an increase of 40%. Even though the temperature rise has been small so far, the expected increases are large enough that future warming is bound to happen. Will it be 2°C or 5°C? I don't know, but I think we should be concerned.

What can we do? Most carbon dioxide of the future will come from the developing world. The United States and the rest of the wealthy world are no longer in control. When you are president, your energy policy needs to reflect this shift. Expensive measures (such as all-electric automobiles) can have a significant impact only if they become cheap enough that the developing world has enough money to adopt them. And even they can fail if they continue their dependence on coal—an electric car that derives its electricity from a coal-burning plant produces more CO_2 per mile than does a gasoline car. Even for "good" solutions, you have to look at the projected costs and the likelihood that those costs could go down. It does no good to set an example if the poorer nations of the world can't afford to follow that example. The best bet, at least for the short term, might be to encourage a worldwide shift from coal to natural gas. That won't stop the CO_2 increase, but it might slow it until other sources of energy become affordable to the developing world.

Next we'll take a broad overview of energy to get a sense of the key issues, the challenges, and the opportunities.

II

THE ENERGY LANDSCAPE

THE ROLE *of energy in world affairs is hard to exaggerate, in part because of the close connection between energy and wealth. Figure II.1 shows the correlation. People in wealthy countries use more energy. Does energy create wealth, or does wealth lead to energy use? It's some of both. It takes energy to run a factory, and wealthy people can afford air-conditioning. Economically developing nations are rapidly increasing their energy consumption for both reasons.*

THE ENERGY landscape is immensely complicated by the fact that fossil fuels—the primary source of energy around the world—are the main culprits blamed for the human-caused (*anthropogenic*) component of global warming. This conflict between wealth and warming is at the heart of the energy debate. The energy landscape is also complicated by widespread misinformation. Despite the beliefs of many people, the United States is not running out of fossil fuel, but only out of conventional oil.

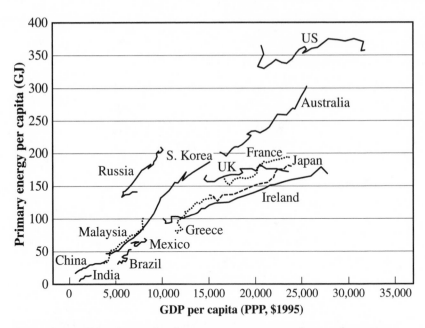

Figure II.1. The relationship between energy use and gross domestic product, per capita. Each country is shown as a squiggly line indicating the ratio over the period 1980–2002. Energy is measured in gigajoules (GJ); 1 GJ is the energy in about 8 gallons of gasoline. The GDP is given in terms of purchasing power parity (PPP), which takes into account the fact that living costs are lower in many countries.

Moreover, the cost of energy is completely out of whack. Here's a dramatic example. The typical cost of electricity purchased from a utility is 10¢ per kilowatt-hour (lower in the southeastern United States, higher in California). Now, instead of using that wall-plug power for a lightbulb, suppose you buy undiscounted AAA batteries from a local fast-food store. I've priced them; they sell for about $1.50 each. (Yes, I know this is expensive, but that was the actual price. They are cheaper online.) How much more does the battery electricity cost compared to wall-plug electricity? Five times as much? Ten times?

No. AAA battery electricity costs *10,000* times as much as the wall-plug electricity! Here's the calculation: One AAA battery delivers about 1 ampere at 1.5 volts for about 1 hour. That's 1.5 watt-hours; for a battery that costs $1.50, that's $1 per watt-hour. A kilowatt-hour (kWh) is a thousand times more, so it will cost $1,000. That's 10,000 times more than the utility price of 10¢ per kilowatt-hour. Yet we do buy such batteries, because they are portable when we need a flashlight we really want it to work. Energy in batteries, which work during power outages, is worth the premium.

The public understanding of energy is so confused that it is difficult to make truly rational energy policy. It will be your job, as president, not only to understand energy yourself, but to explain and convince the public about the relative costs and risks of fossil fuel, alternative energy, nuclear, and energy conservation. As president, you must be the country's energy instructor.

Even though energy is a commodity, you won't find it listed on the commodity pages of the *Wall Street Journal*. You will find energy proxies there—coal, oil, and gas—primarily because of their value in delivering energy. Yet the cost of these energy proxies depends critically on the form in which the energy is delivered, as I illustrated with the AAA battery example. For the same energy delivered, gasoline costs about 2.5 times as much as retail natural gas, and about 7 times as much as wholesale gas! So why do we continue to use gasoline? Why don't we switch to cheaper coal? The answer is something all future presidents and savvy investors need to understand: because our automobile infrastructure (factories, filling stations, delivery systems) was developed over the past 100 years to deliver gasoline, and this inherited infrastructure is too extensive to change rapidly when the price of oil suddenly skyrockets. We are stuck with what economists call an "inefficient market" in energy. The price discrepancy also suggests a big future for energy conversion, and indeed there is a great deal of investment currently going into the required technologies.

Table II.1 shows the cost of energy. The heating column assumes that all the energy in the fuel is converted to heat, meaning 100% efficiency. The last column assumes that the heat is used to generate electricity. Large power plants are relatively efficient, typically

Table II.1. The cost of energy. The entry for coal is the price for anthracite; other coals are cheaper. The cost of natural gas has been changing, and in 2012 dropped to $2 per thousand cubic feet.

Fuel	Cost of fuel	Cost per kWh used for heating	Cost per kWh of electricity
Coal	$60 per ton	0.6¢	2¢
Natural gas	$4 per thousand cubic feet	1.4¢	4¢
Gasoline[a]	$3.50 per gallon	10¢	50¢
Electricity to home[b]	10¢ per kWh	10¢	10¢
AAA battery	$1.50 each	$1,000	$1,000

[a]For gasoline, assumed internal combustion efficiency is 20%.
[b]For power plant electricity generation, assumed efficiency is 35%.

35%–50%, but automobile efficiency is lower, typically 20% or less. These numbers change as the prices fluctuate, and the table includes some retail and some wholesale prices, but it gives a sense of the enormous disparities in the energy market.

Some people think energy is available "for free" from solar radiation, wind, and geothermal (heat coming from the hot interior of the Earth). But of course, such energy is no more free than is energy from coal. The cost for coal is in the mining, processing, and delivery. Solar has similar costs; it is "mined" using solar cells, processed using electronic devices called inverters and transformers, and delivered by the power grid. At present, the sum of these costs is higher for solar than for coal, for the same energy delivered.

In addition, coal is not necessarily cheap if you include its environmental consequences—everything from the traditional pollutants (soot, sulfur dioxide, mercury, nitrous oxides) to the controversially new "pollutant" carbon dioxide (so defined in a 2010 US court decision). The costs of the pollution are borne not by the energy user but by the rest of the world, so there is little economic incentive for the consumer to switch to more expensive energy sources.

Recycled Energy

One of the most remarkable insights about energy is that there is a source that is typically cheaper than coal, and yet often neglected by non-experts such as home owners. This source is energy that is reused—recycled energy, conserved energy. An example is energy used to heat a house that is trapped in the house, not allowed to escape out the windows or through the walls. Keeping it in costs money—but more frequently than you might guess, the cost of improved insulation is lower than the cost of a few years of extra energy.

Energy conservation was given a bad name by President Jimmy Carter when, during the oil embargo of 1979, he declared the crisis to be the "moral equivalent of war." Among other things, Carter urged US citizens to turn down their thermostats in winter—and to put on a sweater instead. True, the hardship was minor; a sweater is no big deal. But many people felt that the quality of their home life had to be lessened as a patriotic duty. Once the crisis was over the thermostats went right up. Carter's call to nationalism had an inadvertent consequence: it convinced US citizens that energy conservation meant enduring discomfort.

Perhaps colder homes in winter were necessary—the crisis was immediate—but an opportunity was lost. President Carter could have told people to put a sweater on, temporarily, but let the US government give you zero-interest loans to install insulation. Then you could turn up the thermostat to whatever temperature you wanted. And the same trick works in summer; insulation to keep the heat out lowers the cost of air-conditioning.

The cheapest form of energy is indeed energy that is not used. David Goldstein, one of the great innovators in conservation, calls it "invisible energy." Amory Lovins, another great innovator, calls it "negawatts." If you can make your heater or your refrigerator or your air conditioner more efficient, then you get the same benefit with less energy. We will talk more about this subject in Chapter 7.

Unlike many other commodities, energy is *not* cheaply stored once it is produced. You can store it in batteries, but batteries are expensive to manufacture and replace after their limited lifetimes. You can store energy by separating hydrogen from water (electrolysis), but the process is inefficient and expensive. I'll devote all of Chapter 10 to energy storage.

Energy Security

The two largest issues in the energy landscape are energy security and climate change. The challenge is to address both of these in reasonable and balanced ways, and much of that involves educating the public on the difference between effective policy and feel-good policy. Some approaches, such as energy conservation, address both security and climate change. Some, such as the conversion of coal to oil, are concerned primarily with energy security. Some, such as large-scale adoption of solar power, are concerned primarily with the dangers of climate change. Liberals tend to worry more about climate change, and conservatives tend to worry about energy security, but as president you will need to address both.

To understand energy security, you must appreciate the enormity of our energy "flow"—the amount we use every day or every year. For example, here are the fossil fuel amounts used by the average US citizen, including industrial use:

- Coal: 18 pounds per day per person
- Oil: 16 pounds per day per person
- Natural gas: 10 pounds per day per person

It is interesting that we use similar amounts, by weight, of all three fossil fuels.

For the United States, the total flow for the entire population is charted in Figure II.2. The largest energy sources are petroleum, coal, and natural gas—the "fossil" fuels made mostly from plant life

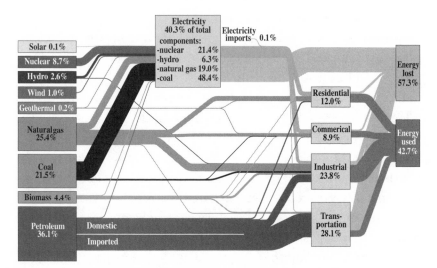

Figure II.2. US energy flow. Sources are on the left, consumption on the right. The total is 3,500 gigawatts, equivalent to 3,500 large power plants. Energy lost refers to unused waste heat. (Based on a Lawrence Livermore 2010 chart.)

(not dinosaurs!) that lived 300 million years ago, when the Earth was hot, atmospheric carbon dioxide was 3–5 times higher, and plant life was far more abundant than it is now.

The most important fact of energy flow isn't the breakdown; it is the enormous size of the flow, about 3,500 gigawatts. That's 3,500 large generating plants; it is 12 kilowatts per person.[1] Generating this much power takes about 300 tons of fossil fuel every second. Three hundred tons per second! Assume, for the purposes of illustration, that all this energy came solely from petroleum. Then the amount of petroleum needed would be about one cubic mile per year.[2]

This image is worth pondering. The United States uses about a cubic mile of fossil fuel every year. Visualize it. Any proposed alternative-energy sources must cope with this enormity.

Moreover, as far as energy is concerned, we live a hand-to-mouth existence. We extract what we need and use it almost immediately. We do put some aside, but only a little. In the United States we maintain a *Strategic Petroleum Reserve* consisting of oil pumped into geologic caverns created in underground salt deposits in Texas and Louisiana. In many ways, the Strategic Petroleum Reserve was a great investment. That oil was purchased at $20 per barrel, and the price of oil has skyrocketed since. The caverns can hold up to 727 million barrels of oil, and are currently almost full. That sounds like a lot, but we are a big country. In the past decade we have been importing over 9 million barrels of oil each day, so if the strategic reserves had to replace imports, the full reserves would be gone in less than 2 months. But there's another issue: we have limited pumping capability. At present, we can extract only 4.4 million barrels of oil per day from this reserve. So if we were suddenly cut off from all imports, we would still have to cut our daily use enormously.

It is worth learning these approximate US numbers: 3,500 gigawatts, 300 tons per second, a cubic mile of oil equivalent per year. They illustrate the huge size of the problem. One large coal or nuclear plant produces 1 gigawatt, 1/3500 of the US need. The enormity of the flow limits the choices for significant alternative fuels. If someone suggests, for example, that we use discarded cooking oil to address our energy problems, you will think, "Do we really throw away 300 tons of cooking oil per second? Do we generate a cubic mile per year?" The answers, of course, are no! and no! The amount of available cooking oil is minuscule compared to our average fuel use. It can help, but only in a minuscule way.

There is a similarly important fact for energy supply: the rapid emergence and rising energy use of the developing countries, particularly China and India. China is so hungry for liquid energy that it is the most powerful driving force for increased oil prices. Whenever the *margin of spare capacity*—the amount of oil that could be pumped around the world minus the amount that actually is pumped—drops below a few percent, the price of oil soars. That's because companies that need oil sign contracts to buy it at premium prices rather than taking the risk that they won't be able to get it. When this happens, news media sometimes report that "speculators" are driving up the price, but that is a gross oversimplification. It is the continuing growth of the economies of the developing world that keeps the spare capacity low, and therefore the price of oil high. The 2007–08 tripling of oil prices took place when the margin of spare capacity dropped below 2%, thanks primarily to China's rapidly rising demand (the Chinese economy has been growing at 10% per year for the last 20 years). Oil prices rose 10% in response to the revolt in Libya because even the low production of that country (under 2 million barrels per day) could affect the spare capacity.

The margin of spare capacity will increase significantly if we build factories to manufacture diesel fuel and gasoline—synfuel—from coal and natural gas. Such synfuel factories could provide an extra measure of supply security, in addition to that provided by the Strategic Petroleum Reserve. The spare capacity will also increase if we succeed in exploiting our recently recognized shale oil reserves.

For every energy technology we have to be sensitive to the differences between the developed world and the developing. Here's an example: the cost of producing solar cells is dropping so rapidly that, I expect, in the near future the cost of the cells will be negligibly small. (There is an analogy in nuclear power: the cost of raw uranium, only 0.2¢ per kilowatt-hour delivered, is virtually negligible.) Yet the cost of solar-cell power in the United States, if installation and maintenance are included, may have trouble competing with natural gas. Natural gas is the primary enemy (economically speaking) of US solar. But the conclusion is very different for China and the rest of the developing world, for the simple reason that labor for installation and maintenance is cheaper than in the United States, and that difference could make future solar cheaper in developing countries, enabling it to compare with natural gas.

4

The Natural-Gas Windfall

THE WORD *windfall* gets its name from the forest. After a big wind, it is easy to gather wood that has been blown off high branches and can be picked up with little effort. The original windfall was easy, cheap energy.

The most important new development in the energy landscape is the recognition that enormous reserves of natural gas trapped in shale, a kind of sedimentary rock, are recoverable. It's a huge new windfall. Although the presence of this gas has been long known, economical means of extracting it have only recently been developed. The exploitability of these shale gases is the most important new fact for future US energy security—and for global warming—and it will have a major impact on economic and political decisions in the next few years and decades.

In 2001, the proven natural-gas reserves in the United States, according to the Department of Energy, were only 192 trillion cubic feet (Tcf). Since we extracted about 20–24 Tcf every year since then, by 2010 it all should have been used up. But in fact, by 2010 the reserves had *increased* to about 300 Tcf. And in 2011, just one year later, the US Energy Information Administration estimated the reserves to be 862 Tcf. Gas experts that I've spoken to believe the actual number is closer to 3,000 Tcf; some think it is even greater. Windfall seems too mild an analogy. It's more like the shmoos in the old comic strip *Li'l Abner*: the more you consume, the more you have.

How can that be? The answer is relatively simple: when making estimates, the US Department of Energy feels it needs to be conservative, to have very high standards to classify a reserve as *proven*. Gas companies have different standards; they just want a supply to be a good bet. They figure out which of their potential reserves are the most likely to be productive; the best of these are drilled. Only the ones that are discovered to have a recoverable supply are then classified by the Department of Energy as proven.

In 1966, 1.6% of US natural gas was produced from shale. By 2005, the number had grown to 4%. By 2011, the fraction was 23%. Now it is about 30% of the total US gas production. That's incredible growth. It was—and still is—a gas rush, comparable in both excitement and potential wealth to the gold rushes of yore. The *New York Times* pronounced, "There's Gas in Them There Hills." Something revolutionary is happening. Figure II.3 shows the spectacular history of this growth.

The new reserves are enormous, and they are transforming not only the energy landscape but also world politics. Docks in Texas

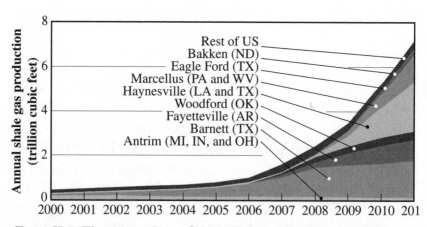

Figure II.3. *The spectacular growth of US shale gas production. The names indicate the geologic formations and the states where they are found.*

and California, constructed to import natural gas, are now being reconfigured to export. Europe is desperately seeking such gas to make itself less dependent on imports from Russia. There appear to be huge reserves in France. And everybody is redoing their geologic surveys.

So much natural gas is available that some industry experts are convinced the wellhead price will remain low, at or below $4 per thousand cubic feet for the next decade or two. As I write this (early 2012), the price is $2.50. (You can check the latest at http://205.254.135.7/naturalgas/weekly/#jm-prices.) To the consumer the cost is closer to $12 per thousand cubic feet, but that might drop too. For that price you can get typically 3.4 gallons of gasoline, but natural gas delivers 2.5 times more energy.

Why aren't we switching to natural gas? Many are. Several of our largest power companies have begun to replace coal power plants with natural gas. Automobiles that run on gasoline can easily be converted to use natural gas with their existing engines. The first to switch over have been the truck and taxicab drivers, who are very sensitive to the cost of fuel. About 130,000 trucks and cabs in the United States have already been converted. The developing world is even more price-sensitive than the United States, and over 7 million vehicles in India, China, Brazil, and Argentina are now using natural gas instead of gasoline or diesel fuel. They can't afford the high-priced stuff.

Our energy infrastructure is huge, however, and it will take time. Natural gas is less compact than gasoline; even compressed, it takes 3 times more space. For that reason, large vehicles such as trucks and buses are the earliest candidates for conversion. Compressed natural gas does have 10 times more energy per gallon than lithium-ion batteries, so it is the real competitor to all-electric vehicles.

Natural gas will be our primary "alternative fuel" for the next few years (or maybe decades), and its competition worries oil producers. Saudi Arabian Prince Al-Waleed bin Talal (Figure II.4) said in May 2011 that he was anxious to increase his flow of oil in order to keep its price down.[3] In the past, the Saudis had claimed they wanted

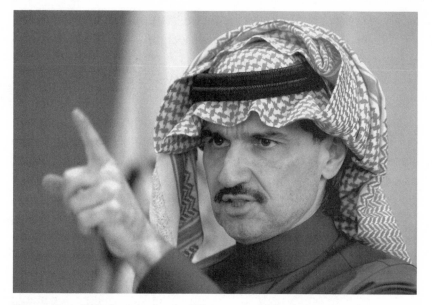

Figure II.4. "We don't want the West to go and find alternatives . . .";
Saudi Prince Al-Waleed bin Talal, speaking at a news conference on
March 9, 2011.

such increases in order to keep the economy of the West vigorous, but this time the prince was more candid. He said (perhaps inadvertently breaking Saudi security), "We don't want the West to go and find alternatives, because, clearly, the higher the price of oil goes, the more they have incentives to go and find alternatives." The danger to Saudi Arabia is that we will develop infrastructure that enables the use of other energy sources. From the Saudi perspective, it is better to keep oil prices low so that the oil-poor world doesn't develop the means to use the competitive fuels.

NATURAL gas, consisting mostly of methane, provides about one-quarter of US energy. It is that smelly stuff that we use to cook food on home gas stoves. It's usually called simply *gas*—although Americans also use that nickname for gasoline. (The British avoid confusion by calling gasoline *petrol*.) Truly natural gas has no odor whatsoever.

That makes it dangerous if a stove is accidentally left on, so the utilities add a little mercaptan to create the rotten-cabbage stink.

Natural gas is the nemesis of underground coal miners; my grandpa was one in Pennsylvania. Natural gas was found adsorbed in the pores of coal, and if a pocket was released into the mine, it would kill by either asphyxiation or explosion. The famed canaries used in such mines were early sensors for this dangerous gas, as well as for carbon monoxide. We still extract natural gas from coal today, mostly from deeply buried coal that is too thin to dig up. Most of the coal bed gas is released by pumping pressurized water down pipes to crack the coal; this is the same method that is now allowing natural-gas extraction from shale.

We once lit our cities and homes with an unnatural gas called "town gas," made by reacting coal with water. Town gas consisted mostly of hydrogen and highly poisonous carbon monoxide. When large sources of methane were discovered, this "natural" gas provided a safer and cheaper alternative. The term *natural* was partly a marketing ploy; it made it sound less dangerous in the home (which it was).

When oil was discovered in Pennsylvania and then in Texas, natural gas was a by-product. It had been dissolved in the underground oil, and it came out as the oil rose to the surface and was depressurized. This "wet gas" was mostly a nuisance to the oil well companies. It couldn't be transported in trucks or trains (and technology for liquefying it didn't yet exist) so most of it was *flared*—burned—at the wellhead. In parts of the developing world this practice continues today, as shown in Figure II.5. Flaring is now discouraged, since it adds the greenhouse gas carbon dioxide to the atmosphere, and a US satellite has been launched specifically to detect flaring around the world. Images from this satellite show extensive flaring in the fields of Nigeria. Flaring makes economic sense to the owners of these fields, but it is particularly cruel, since energy-deprived people live close enough to the flare sites to see the flames. About 5% of all the natural gas produced worldwide is currently flared.

Flaring is now illegal in many countries. There are also good eco-

Figure II.5. A natural-gas flare in Thailand.

nomic reasons to avoid flaring: The natural gas can be pumped back down into the well, where it helps push out additional oil, resulting in "enhanced oil recovery" and additional profit. At worst, pumping the gas back down simply stores it for future sale. And the gas can also be liquefied by cooling to below −162°C (−259°F). Such cooling reduces the volume of the gas by a factor of 750, allowing huge quantities to be carried in refrigerated tankers. New super-duper tankers developed by Qatar can carry over 100 kilotons of natural gas.

Some people worry that such huge tankers are potential targets of terrorists. In fact, natural gas by itself doesn't explode; it has to be mixed at just the right level (5%–15%) with air, and that's not easy to do. However, there is a danger that some of the liquefied gas could come in contact with water (perhaps from a terrorist bomb), and then could suddenly warm and turn to gas (diffidently called a "rapid phase transition" by the experts). The gaseous form takes 750 times more volume than the liquid, and the rapid expansion is a "physical," or "cold," explosion that can destroy more of the tanker and release more liquid natural gas. If the worst-case scenario were to occur for the largest tanker in the world, and a runaway explosive turned all the liquid to gas, the energy release would be about 1 kiloton equivalent of TNT. That's about the same energy released (mostly from burning jet fuel) when terrorists crashed two airplanes into the World Trade Center in 2001.

Fracking and Horizontal Drilling

Hydraulic fracturing, known as *fracking*, is one of the two key technologies for shale gas. It had been known for over 50 years but not widely used. The method was improved in the 1980s by oilman George Mitchell and applied to the Barnett Shale, a rock formation in Texas. It took a decade to make the method profitable. Fracking found its true value in 2002 when Devon Energy Corporation combined it with a second technology, called *horizontal drilling*. In the following years, tens of thousands of wells were drilled in the Fayetteville Shale of Arkansas, the Woodford Shale of Oklahoma, and the Haynesville Shale of Mississippi and Louisiana. These developments attracted little national attention until a company called Range Resources began developing the Marcellus Shale in the northeastern United States.

The current gas rush truly is comparable to the gold rush of 1849. Those landowners who have retained their mineral rights receive 12.5%–20% of the value of the gas produced from their lands. The Barnett Shale in north Texas produces from an area of about 7,000 square miles and has made thousands of landowners into millionaires.

FRACKING is simple in principle. Drill into the shale layer and then send down high-pressure water. If the pressure provides a greater force than the weight of the rock above, the rock will crack to let in the water. That cracking allows gases adsorbed on internal pores to escape. Release the pressure, and the water comes streaming back, along with the gas.

You can do better. In addition to water, send down sand or small pellets—tough spheres that will settle into the cracks. Then, when the pressure is dropped, the cracks remain open, allowing gas from remote cracks to continue to diffuse toward the well and be extracted.

Some opposition to shale gas extraction comes from people who live close to the proposed well, who fear damage to the local environment. The problem is that the water used to fracture the rock comes out of the mine polluted not only with salt, rock, and mud, but with many chemicals (hundreds of different ones) that are introduced with the water to make the process work better. In the past, much of this water was simply dumped into the local watershed, leading to local but severe pollution of streams and rivers. (Fracking was virtually unregulated, thanks to an exemption to the Safe Drinking Water Act.) Moreover, the industry refused to identify all the chemicals, partially to protect its trade secrets, but perhaps also for fear of additional regulation. There already have been accidents. In 2011, an eruption at a gas well in Pennsylvania released thousands—possibly tens of thousands—of gallons of polluted fracking fluid over the berms meant to retain it, forcing seven nearby families to evacuate their homes and threatening pollution of local trout streams.

Other opposition to shale gas extractions comes from environmentalists who worry about natural gas leaking into the atmosphere; as a greenhouse gas it is 23 times more potent than carbon dioxide. Nominally because of all these problems, in 2011 France instituted a moratorium on fracking, even though it imports 95% of its natural gas.

Can the fracking pollution problem be solved? Yes, easily. When we think of technical issues involved with future energy, from the technology of solar to the storage of nuclear waste to carbon capture and sequestration, certainly eliminating the pollution of the fracking water must rank among the easiest. It is straightforward but expensive. There are many approaches, including scrubbing the water before it is released and reusing it in nearby or even distant mines.

Companies have a financial incentive not to spend money unless their competition also has to spend money; that means the solution to fracking pollution is regulation. Requiring that only clean water be released will also inspire competition among those who can do the cleaning, so regulation could eventually reduce the cost. Accidents will occasionally happen, but as with airplane crashes, the key is transparency so that we can learn from the mistakes and minimize recurrences.

It is always worthwhile to look at the numbers. The accidental spill in Pennsylvania was described in newspapers as releasing thousands to tens of thousands of gallons of filthy fracking fluid. For ease of calculation, let's assume that the actual number was 7,500 gallons. That's 1,000 cubic feet—the volume of a cube 10 feet on each side. It can certainly do damage, but it is not a stupendous amount; it is less than the capacity of a single railway tank, and it could be handled with moderate facilities.

In addition to fracking, the other essential technology for shale gas extraction is horizontal drilling. Horizontal drilling was first developed for oil wells. It allowed the industry to drill straight down at one location, and then, as the pipe advanced, to "bend" the hole and the pipe so that they would continue to drill in a different direction. In addition, after one well had been drilled, the drill could be sent back down and, once at the bottom, go off in a different direction. As shown in Figure II.6, the same vertical hole can be used to reach a large number of different oil deposits, either sequentially or simultaneously (by using branching pipes). This approach significantly reduces the cost of drilling, particularly for very deep deposits. And if one location proves dry, you can use the same vertical hole to probe others.

For shale gas to be practical, horizontal drilling is essential. The reason is that the layers are typically thin, only a few tens of meters deep, yet the shale can extend thousands of meters in the horizontal direction. Traditional hydraulic fracturing cracks the rock only close to the pipe, so drainage of gas is much more effective if

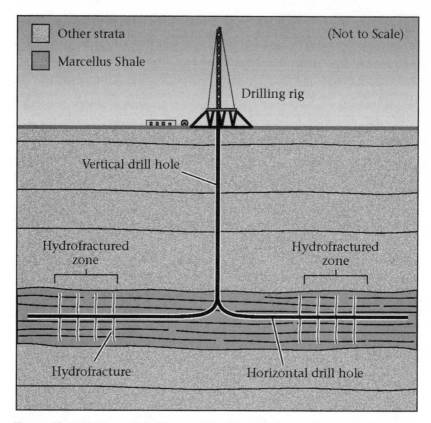

Figure II.6. Horizontal drilling and fracking enables wells to reach gas trapped in thin but extensive layers of shale.

the horizontal pipe and its associated fractures extend for a thousand feet or so through the entire formation. Only when horizontal drilling was proven practical for other uses did shale gas become a viable target.

Shale Gas Reserves

The map in Figure II.7 shows the regions of the United States that are now believed to have economically recoverable gas from shale. The area is enormous. Recall that the Energy Information Administration estimates 862 Tcf (trillion cubic feet) of natural gas in these areas, but that some gas experts think 3,000 Tcf is a better estimate. For comparison, the huge conventional gas reserves in the Middle East are about 5,000 Tcf. Qatar, the tiny country in the Persian Gulf, has about 1,000 Tcf.

Figure II.8 shows the shale gas reserves estimated by the US Energy Information Administration for 32 countries. According to this survey, the total in the countries mapped is 6,622 Tcf, of which 13% is in the United States.

Many of the reserves are found in the countries that truly need them. Europe hopes to wean itself from its current dependence on Russia. France currently imports 95% of its gas, but it has "technically recoverable shale gas reserves" sufficient to last for 100 years[4] (provided the French government reverses its ban). China, the world's largest consumer of fossil fuels, primarily coal, has 1,275 trillion cubic feet according to the EIA estimate. That's enough to cover its current natural-gas consumption rate for 400 years.

If countries switched from coal to natural gas, these reserves could make a huge difference in the world production of carbon dioxide.

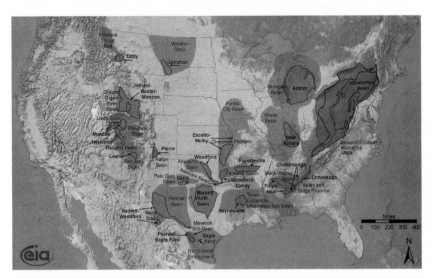

Figure II.7. Shale gas regions in the continental United States.

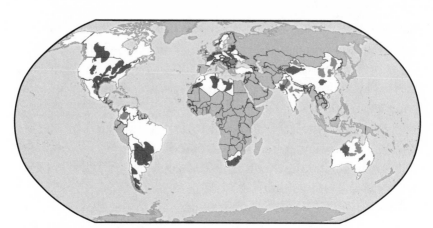

Figure II.8. The EIA's map of 48 shale gas basins in 32 countries. Only reserves in the countries indicated in white are included.

Natural gas produces only half the greenhouse gases that coal does, for the same energy. Moreover, its local pollutants (mercury, sulfur, carbon particles) are much lower. Natural-gas exploitation in China could be very attractive for local health reasons and, in addition, would significantly lower Chinese CO_2 emissions if the country switched over.

All these reserves may sound big, but compared to yet another source, you ain't heard nothin' yet.

Ocean Methane

Deep in the oceans there is a source of natural gas that is far more extensive than shale gas. These icy deposits full of methane are found on continental shelves. Figure II.9 shows what this fantastic material looks like. It contains more water than methane (5 times as much), but it burns anyway. This substance is called a *methane hydrate*, or alternately, *clathrate*.

Methane hydrate forms when methane seeps up from the sediment at the bottom of the ocean and comes in contact with cold water. The water at the bottom is typically 4°C (39°F), just barely above freezing.[5] But if methane mixes with such water, the methane molecule seeds the growth of a water crystal cage around it, and the combination is solid. It isn't ice, but it is very much like ice. It forms only in cold conditions under very high pressure, typically 50 times greater than atmospheric. For that kind of pressure, the depth underwater must be at least 1,500 feet.

Nobody knows how large the methane hydrate deposits really are, but "conservative" estimates suggest that they *exceed* those of shale gas by a factor of 10! The US Department of Energy has said that the reserves could be even 100 times greater than those of shale gas. Figure II.10 shows the US Geological Survey's 1996 map of methane hydrate sites. They are found primarily along the coasts, on the continental shelves. We don't know why this methane is there. It does not appear to be associated with fossil carbon. It may come from

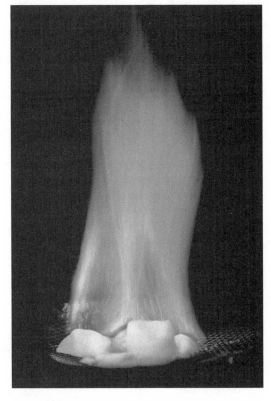

Figure II.9. Methane hydrate looks like ice, but it can be ignited to burn in air.

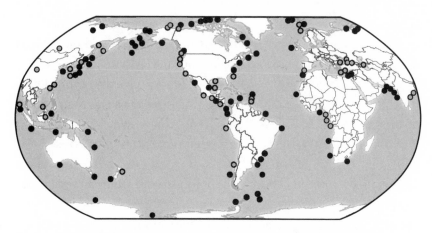

Figure II.10. Methane hydrate locations determined by the US Geological Survey. The deposits are found primarily on the continental shelves.

bacteria living on the shelves, or perhaps it is the result of primordial methane from the formation of the Earth gradually seeping up from deep within.

You might have heard about methane hydrate in the news during the Gulf oil spill. BP maneuvered a funnel above the gushing oil to try to collect the oil, but the oil contained methane, and when the methane came in contact with the cold water at the bottom of the Gulf it combined to form hydrate, like the material shown in Figure II.9, and that solid clogged the funnel, causing this initial attempt at collecting the leaking oil to fail.

Can the vast deposits of hydrates deep underwater be harvested? Many experts say no, but they may be wrong. Fifteen hundred feet may sound deep, but oil rigs in the Gulf already operate in 5,000 feet of water, and then drill several miles into the rock under the seafloor. Automated machines can reach these areas and send warm gas or water down to soften the hydrates and release the methane, which could then be collected in funnels. Or the hydrates could be crushed and brought up on conveyor belts; on the way up, the hydrates would be warmed in the shallower water and the methane would be released. Harvesting hydrates isn't trivial, particularly since most of the deposits are not made of pure crystals like the one in Figure

II.9, but are mixed with clay, leaving a hydrate concentration of only a few percent. Recovery is also made difficult by the corrosive properties of salt water. But recovering the methane does not seem impossible. One cubic foot of methane, on the Earth's surface, contains 1,000 BTUs of energy; that's enough to raise the temperature of 1,000 pounds of water by 1°F. (In fact, that is the definition of the BTU—the British thermal unit.) So if supplied with air or oxygen, the methane deposits have plenty of energy to release themselves.

The great danger in exploiting the undersea reserve comes from the potential global warming, not only from the carbon dioxide created when the methane is burned, but from the danger of leaked methane—a gas with 23 times greater greenhouse effect than CO_2. Some scientists have speculated that a runaway release of undersea hydrates was responsible for the great geologic catastrophe called the Permian-Triassic extinction, an event that took place about 250 million years ago and that rendered extinct 96% of all marine species.

Nobody I know of is presently planning to exploit the undersea methane reserves, in part because there is plenty of shale gas that is cheaper to get. By the time we're ready to mine methane hydrate, we should know a lot more about the amount of greenhouse warming that results from human carbon dioxide. For now, I think of the undersea methane hydrates as a potential long-term source for the distant future.

5

LIQUID ENERGY SECURITY

THE ENERGY crisis . . . What energy crisis? The United States has huge reserves of coal, enough to last for at least a century. We have enormous reserves of natural gas and oil in shale rock. We also have lots of sun and wind. We're running low on cheap uranium, used for nuclear power, but the cost of uranium ore is such a tiny part of the cost of electricity (about 2%) that we can tolerate a big price increase, and that means we will not run out of affordable uranium, not for centuries.

We don't have an energy crisis; we have a *transportation fuel* crisis. We don't have an energy shortage; we have an *oil* shortage. We're not running low on fossil fuels; we're running low on *liquid fuels*. Look at the three charts in Figure II.11. The first one shows the petroleum reserves of various countries. The largest reserves—the tallest bar—are in Saudi Arabia, the king of oil. And the pathetic little bar on the far left represents the dwindling US oil supplies.

The second chart includes oil as the lower part of each bar, but on top of each bar I added in the reserves of natural gas. I converted the amount of natural gas into "barrels of oil equivalent"; that is, each bar represents the total amount of energy delivered by each

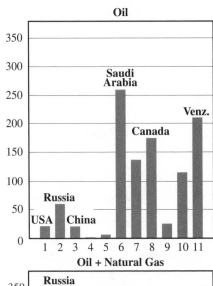

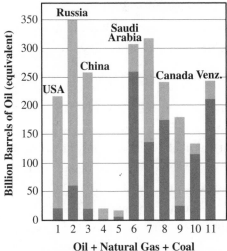

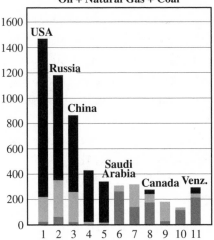

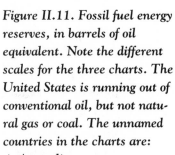

Figure II.11. Fossil fuel energy reserves, in barrels of oil equivalent. Note the different scales for the three charts. The United States is running out of conventional oil, but not natural gas or coal. The unnamed countries in the charts are:

4. Australia

5. India

7. Iran

9. Qatar

10. Iraq

commodity. Now you'll see that the United States has about 70% as much energy available in fossil fuels as has Saudi Arabia.

In the third plot I changed scales, so all the oil and natural gas is indicated now by shorter bars, but I added in our coal reserves. Look at the numbers. The United States has 1,470 billion barrels of oil equivalent in fossil reserves, nearly 5 times as much as the 307 of Saudi Arabia. The United States is the king of fossil fuels! Poor Saudi Arabia.

Is that good news or bad news? It depends on whether you're concerned more about energy security or about global warming.

GASOLINE, *diesel fuel, jet fuel, oil, petroleum*—for this discussion they are synonymous. Petroleum was once called *rock oil*; it is the version of oil that comes from the ground (in contrast to olive oil and whale oil) and that we convert and refine into gasoline, diesel fuel, fertilizer, plastics, and asphalt.

This is not the first time we've run out of oil. In the mid-1800s, the best form of lighting for homes and businesses was whale oil, and the best whales were being decimated. At the time, this problem was considered not a whale crisis but an oil crisis. The supply of whale oil peaked in 1845 at 15,000 gallons per year, and then began to drop; as a consequence, the price shot up, and by 1852 it had doubled. But the discovery of abundant rock oil in Pennsylvania in 1859 provided an urgently needed substitute. By the end of the year, rock oil production in Pennsylvania already exceeded that of whale oil at its peak. The major use of the new oil was to make kerosene for lamps.

The oil rush was at hand. We were in the midst of an energy revolution. John D. Rockefeller made his oil fortune on kerosene, before the automobile came about. In fact, it was this discovery of petroleum that made the internal combustion engine possible, and that led directly to the automobile and the airplane.

Was it merely luck that petroleum discoveries were made just in time to replace the declining whale oil? In fact, the discovery in Pennsylvania was made in a region in which natural oil seeps were common. Such rock oil had been used primarily for medicinal purposes. There was no real market for rock oil until the whale oil crisis hit. Whale depletion drove the search for something new. There is a certain irony in the fact that just over a century ago, petroleum was an *alternative* energy source.

Some early cars ran on coal, but they couldn't really compete with gasoline autos. Pound for pound, gasoline delivers over 60% more energy than does coal. Moreover, it leaves the tank empty, not full of ash. Think about the miracle of gasoline. You can pump it into your car (rather than shovel it) in just a few minutes. When you do this, you're transferring energy at the rate of about 4 megawatts—enough to briefly power 4,000 homes.[6] Gasoline prices are quite variable, but let's take a typical US price of $3.50 per gallon. (They may be higher or lower today—the day you're reading this—because the price is more volatile than gasoline itself.) At 35 miles per gallon, it costs you 10¢ worth of fuel to drive a mile. A dime per mile! That's incredibly cheap, especially when you realize you can have five or more people in that car. No wonder we love automobiles.

Energy use is closely tied to the gross domestic product, as we saw in Figure II.1. Energy conservation can break that relationship, but let's put that aside for now; we'll return to it in Chapter 7. We do need to be more efficient, and there are good economic incentives to do that.

Hubbert's Peak

When people in the United States talk about our energy crisis, what they really mean is that we're running out of a domestic supply of the very convenient form of energy we use for transportation: oil, or as the British say, petroleum.

A widely believed way of predicting resources, based on something called Hubbert's peak, has been used to foresee higher and higher oil prices accompanied by dwindling supply. In 1956 Marion King Hubbert, a geoscientist working at Shell Research Labs, stated some obvious facts in such a clear and concise way that his ideas have become central in US government policy. Hubbert pointed out that when a new natural resource is discovered (think of the 1859 discovery of rock oil in Pennsylvania), the initial high price drops as exploration discovers new sources. However, natural resources are finite, and eventually the demand overreaches the supply. Prices then rise as the supply drops. The maximum production is now called *Hubbert's peak*. The United States reached its Hubbert oil peak in the 1970s, and the world right now is close to the time predicted for its oil peak.

Hubbert's analysis is appropriate if there are no substitutes or new sources for the commodity. Recall that we did find rock oil to replace whale oil. The obvious substitutes now for rock oil are natural gas and synfuel.[7] And there is a potential new source. Just as shale gas was made accessible by fracking and horizontal drilling, oil engineers believe they may be able to recover vast new reserves from *oil shales*; I'll discuss these in more detail in Chapter 6.

The oil shortage crisis in the United States is growing. The Department of Energy (DOE) was created by President Jimmy Carter in 1977.[8] His explicit reason was to lessen our dependence on foreign oil, to give us "energy independence." The program initially succeeded; by 1984, imports had dropped by 50%. But lower oil prices encouraged a reversal of this trend, and by 1994, imports exceeded the 1977 peak. Then, for the first time, imports surpassed US domestic production. In 2011, the US imported 3.05 billion barrels of oil at a price averaging $99 per barrel, for a total cost of $302 billion. The US trade deficit that year was $573 billion. That means 53% of our trade deficit in 2011 came from importing crude oil.

Because it is next door, Canada provides about half of our oil imports; and because Canada is a friend, we feel a certain degree of psychological security against another embargo. But it provides no economic security against rising prices of oil and anticipated shortages.

Under President Carter, the newly formed Department of Energy initiated programs in solar and wind, and in technology to convert coal to diesel fuel. All of these programs were terminated under President Ronald Reagan. Some argue it was Reagan's ideology that prompted the shutdowns, but it is also true that dropping oil prices made alternative energy relatively unprofitable, requiring large subsidies. Oil had reached (in constant dollars) $111 per barrel in the Carter years, but then the price dropped during Reagan's administration to $22 per barrel. None of the alternative-energy technologies could compete with such a low price.

The persisting problem with alternative energy is that it has been more expensive than energy from fossil fuel. Saudi Arabia can drill its oil for about $3 per barrel. The historic price for oil, in current dollars, was $20 per barrel—so the Saudis could make a handsome profit. At current barrel prices—$70 to $100 and above—their profits are enormous. The price skyrockets whenever the world demand begins to exceed the supply.

In the last few years, US oil production has begun to rise, thanks largely to improved technology and growth in the number of rigs. In 2011, for the first time since 1977, the United States imported less than it produced. But even if this trend continues, the large imports have dire consequences for our trade deficit.

One potential solution to the liquid fuel problem is synthetic fuel—synfuel. We can make gasoline from coal; that's called CTL for "coal to liquid." We can also make it from natural gas; that's called GTL. The original method for manufacturing synfuel was the Fischer-Tropsch process, invented in the 1920s and effectively used by Nazi Germany in the 1930s and 40s and by South Africa during the apartheid era. Both of these countries had abundant coal but could not get gasoline or diesel fuel. That sounds like the US situation today. So why aren't we making synfuel?

The real problem is uncertainty in the market. According to industry experts that I've spoken to, synfuel costs about $60 per barrel to make. Even now, when the price of oil is well above that, investors fear that the Saudis will drop the price of petroleum after the synfuel plants are in operation, driving the plants to bankruptcy. Recall the drop in prices from $111 to $22 per barrel that took place between 1977 and 1986. The only thing that could prevent Saudi Arabia from undercutting synfuel prices is unfulfilled demand, and if we hadn't gone into a recession, we might be at that point even now. The continued growth of China and India and the rest of the oil-hungry developing world has stretched the production capacity to its limits. That's why the price of oil is high and will very likely remain high in the immediate future.

How high can the price of oil go? In the long term, it should not be able to stay above the synfuel price of $60 per barrel. Only the time delay between the certainty of continued high price and the construction of the synfuel plants will allow the price to rise temporarily. That period of limbo is where we are now, and the Saudis are worried. They are just as addicted to their huge oil revenues as the United States and the West are addicted to oil.

There is another upcoming source of liquid fuel that could drive the price of oil lower, and that could even challenge the profitability of synfuel. It's called shale oil.

6

SHALE OIL

THE FASTEST-BREAKING news in the energy world is not natural gas but something with perhaps even greater potential. It is something that even a year ago I had misplaced in the same category as undersea methane hydrates—as a long-term source for the distant future. But nothing drives technological breakthrough as much as the potential for great profits. The continued high price of oil has inspired innovation in a form of fossil fuel that had been dismissed as impractical even by most experts until recently.

Shale oil, like shale gas, is trapped in rock, inaccessible (we thought) without the expenditure of huge amounts of energy—maybe even more energy than could be recovered. The classic idea was to mine the shale, heat it, let the "tight" oil-related gunk called kerogen ooze out, and then convert the kerogen to diesel and gasoline through the magic of chemistry. This process was called "retorting," in analogy with other laboratory chemistry, and it seemed so expensive that it would have to wait for much higher oil prices—prices that might never come, thanks to synfuel competition. In addition, the waste produced by retorting, the dried-out shale, occupies a larger volume than the rock that was mined, making for a new environmental challenge.

Yet the amounts of shale oil in the United States are huge—

stupendous, actually, with plausible estimates of over 1.5 trillion barrels, 5 times the reserves of Saudi Arabia! Recall that the United States uses about 20 million barrels per day, so 1.5 trillion barrels would last 200 years. But shale oil is thick gooey stuff, embedded in rock, deeply buried. We've known about it for decades, and it looked so unreachable that energy projections typically ignored it. It seemed impossible that there could be some clever, low-energy, cheap way to get to it.

Inspired by the cost of oil, by the success of shale gas horizontal drilling and fracking, and by some smart physics and chemistry and engineering, oil companies including Shell, Chevron, and ExxonMobil have been developing ways to extract the oil, and their quick success has been amazing. Shell's method is called the "In-situ Conversion Process." They heat the deep rock (1–2 kilometers down) using electricity to temperatures of 1,200°F–1,300°F (650°C–700°C)—an approach that seems so wasteful that I probably never would have considered it—and they combine this process with conventional horizontal drilling and fracking. (It's amusing that they now consider this to be *conventional!*) They let the heated rock simmer for 3–4 years. This process is similar to the natural one that converts heavy kerogen into sweet oil (smaller hydrocarbon chains), but the heating speeds it by a millionfold. The kerogen breaks up into smaller hydrocarbons, which move more freely through the fracked rock.

Initial tests of the approach have been surprisingly (to me) successful. The energy of the extracted oil is 3.5 times larger than the energy expended, including that used to heat the rock. That's plenty good enough. Shell estimates that the process costs about $30 per barrel; other companies have stated that the process makes economic sense as long as the price of oil remains above $60 per barrel. The difference in these estimates may lie in the cost of bringing the product to market, and the need to earn a profit on the investment. Or maybe the other companies are simply hiding the huge profits to be made. At $100 per barrel, a new oil-shale well can become profitable in less than a year.

There are, of course, environmental concerns. Should we really be happy about the discovery of a new source of carbon? And there are other worries; for example, what if the heated rock releases the newly mobilized oil into the water table? Shell has developed an "ice wall" technology that involves freezing the rock and soil in the region surrounding the deposit, to prevent this. The company has tested this approach at several locations, but it isn't yet fully proven for large deposits. There is also a wastewater issue, as there is with shale gas, and a water shortage issue; in Texas, an extended drought had made water for fracking a precious resource. Some people argue that the environmental dangers of drilling on land are much less than those drilling out at sea, and that Texas is less vulnerable to oil pollution than are the fragile coast and tundra of Alaska.

According to Shell and the Department of Energy, one square mile can produce a billion barrels of shale oil, and there are thousands of square miles of the right kind of shale under the Colorado Plateau alone. In North Dakota, the production of shale oil at the Bakken field jumped from nearly nothing 4 years ago to 400,000 barrels per day by 2012, and the site is expected to be yielding a million barrels per day soon. Recall that the total imports of the United States have averaged about 10 million barrels per day. The Eagle Ford Formation in Texas is producing 100,000 barrels per day, and that number is expected to rise to 420,000 in the next few years. Oil companies are investing about $25 billion this year in drilling some 5,000 wells. By the end of this decade, some estimate that 25% of US oil consumption will come from these new shale oil fields. It could be more.

Shale oil production could truly be a disruptive technology, with a large and positive impact on the US balance of trade, severe repercussions for the OPEC oil cartel, and a serious challenge to alternative-transportation technologies, particularly natural gas and synfuel. Shale oil could turn conventional oil into the new whale oil, replaced by a far more abundant source. It could be wonderful for US energy security, but potentially disastrous for the environment if concerns about global warming prove valid.

7

ENERGY PRODUCTIVITY

A penny saved is a penny earned.   —*Poor Richard's Almanac*

A gallon saved is a gallon not imported.

—*Proposed for the next edition*

THE BIGGEST item on the alternative-energy menu is not nuclear power, not solar, not natural gas, not coal or oil or synfuel or any other supply. It is something with much greater potential, cheaper than cheap, and it could be ready and in place in time for your reelection campaign. This magical energy source is not actually a new energy source at all; it is an increase in our *energy productivity*. That means accomplishing the same tasks we now do, but using less energy.

The opportunity is enormous, and yet it is hardly being exploited. There is no more urgent energy-related action for you to take when elected than to create and implement a sensible energy productivity policy. The most amazing aspect of your policy will be its cost: zero. Zilch. In fact, it should be profitable. That's what I meant by "cheaper than cheap."

How profitable? Several clear and safe investments that I will describe in some detail will return interest at rates higher than 10%, and for some examples, at a much higher interest rate.

Will it hurt? As I mentioned earlier, in 1979 President Jimmy Carter (Figure II.12) asked the public to turn down their thermostats (it was winter) and "put on a sweater." Most people didn't like

Figure II.12. President Jimmy Carter, wearing a sweater, with no fire in the fireplace.

it. Soon, as the oil supplies increased and the price crashed—in current dollars, from a high of $111 per barrel in December 1979 down to $22 per barrel in 1986—many people took off their sweaters and turned up the heat. Worse, they were left with a strong impression that energy conservation meant a change in lifestyle, to one that they considered less comfortable. That was a very unfortunate outcome, because increasing our energy productivity does not necessarily require any sacrifice at all. Many people (particularly in Berkeley) say that putting on a sweater is *not* a decrease in quality of life. So let me make a stronger statement: properly implemented, an increase in energy productivity does not reduce even the *perceived* lifestyle of those who participate. Moreover, it is probably the biggest step that can readily be taken to reduce the US balance-of-payments deficit. And unlike synfuel, shale gas, and shale oil, it does not endanger the environment.

All this sounds incredible for several reasons. The first is the old dictum that there's no such thing as a free lunch. How can there be an action with no downside? The second is that people aren't stupid. If this is such a great idea, why isn't it being done already? There are explanations for both of these, but they lie primarily in the realm of psychology—the result of poorly explained and poorly implemented policies in the past.

Let me begin by giving a simple example that shows in a dramatic way what opportunities are being missed.

Invest for a 17.8% Annual Return, Tax-Free, with No Risk

No, this is not the Muller Ponzi scheme. Bear with me while I work out the numbers. Because the 17.8%-per-year return is tax-free, it could boost the effective return as high as 27%, depending on your tax bracket. Not only that, but it is completely hedged against inflation, particularly the likely inflation in the cost of energy. Moreover, the investment is *patriotic*. It helps reduce both the US balance-of-payments deficit and our dependence on foreign oil. Too good to be true? No—too mundane to be taken seriously.

Here's the secret investment opportunity: install insulation in your attic.

Installing insulation may not yield high returns in a *new* house—and not in yours if you already have good insulation—but Art Rosenfeld, a former official in the Department of Energy, estimates that half of the older homes in the United States would benefit from added insulation.

Now you may have lost interest. Argh! It's the very boring nature of this investment that may explain why people don't take advantage of it. It turns out that ordinary boring but comfortable energy conservation yields a far better return on investment than virtually anything else, and it is essentially risk-free. The only "danger" is that energy prices might plummet; such a drop is highly unlikely, but if it does happen, you probably will be happy anyway. On the flip side, if energy prices skyrocket, your profit is even greater.

How does this incredible investment really work? Let's start with an example adapted from a US government web page:

www.energysavers.gov/your_home/insulation_airsealing/
index.cfm/mytopic=11360

The page gives a formula for calculating the *payback period*—that is, the number of years it takes before the savings in energy pay for the cost of installing insulation. Without going into the details, let me simply say that the formula takes into account plausible costs, plausible efficiencies, plausible everything, and shows in its example a payback period of 5.62 years. Check the page for the details.

Surveys taken by public utilities have shown that many people think a period of 5.62 years is awful; it takes way too long to get your money back. It sounds like you're making a loan (to your home) and receiving no interest. After 5.62 years, you're just even. Terrible investment! They conclude they can't afford to install insulation. But let's look at the investment more carefully.

For clarity, let's put in some numbers. Assume that you spend $1,000 on insulation for your attic. Then, according to the numbers worked out in the example, after 5.62 years you would have saved $1,000 in energy cost.

Now let me try to convince you that doing that is actually a wonderful investment. So let's return to the Ponzi scheme example. Suppose instead of putting your cash into insulation, you had invested it with Bernard Madoff. Let's assume his scheme was actually legitimate, not crooked. He was offering 11% per year, so after 5.62 years (assuming you didn't compound), you would have earned 55%—a gain of $550. That kind of rate was considered spectacular, particularly because it appeared safe (Madoff didn't have big fluctuations in his interest rate; it was a steady 11%).

Yet, if you had spent that $1,000 on insulation instead of handing it over to Madoff, at the end of the payback period you would have had $1,000 extra in your bank rather than $550. It would be the money you didn't have to spend on heating or cooling. This is real money in your bank account.

Ah, you object—in Madoff's scheme you could also have recovered your principal, the original $1,000. (We'll ignore the fact that the Madoff scheme was a scam.) Remember that when you put the money into your house, you increased its value. That payback may not be liquid, but it is real. You haven't lost the $1,000; you will get it back (and maybe even more) when you sell.

What if you never sell? Then you can't extract your capital, at least not in the same way you can withdraw money from a bank or sell stock. But why would you want to? The capital continues to pay 17.8% per year—in the form of reduced heating and cooling costs—forever. You would never want to extract capital from such a great investment. If you had that rate of return from a stock, and it was secure and tax-free, you'd be crazy to withdraw it.

The problem is that the public doesn't understand the relationship between payback period and effective interest rate. The simple equation is this:

$$\text{Interest rate} = \frac{100\%}{\text{Payback period}}$$

So a payback of 5.62 years gives an interest rate of $100\% \div 5.62 = 17.8\%$. This equation does have to be modified if the value of the thing purchased does not constantly add to the value of what you own. In that case, a depreciation number must be included in the equation:

$$\text{Interest rate} = \frac{100\%}{\text{Payback period}} - \text{Depreciation rate}$$

(If inflation or deflation becomes significant, it must also be included.) I'll assume that insulation does not depreciate significantly, but some objects (for example, compact fluorescent lightbulbs) have a finite lifetime and do depreciate.

Suppose the cost of energy goes up. If that happens, then the interest rate is even higher. What if utility bills go down? Suppose they're cut in half. (That's not a very plausible scenario, but it is possible.) Then the effective interest rate is also cut in half, down to 8.9% instead of 17.8%. Even if the utility rates are cut by 75% (truly implausible), your interest rate is still 4.45%. Pretty safe investment, don't you agree?

Moreover, if you got this return from a bank or from stocks, you would have to pay taxes on it. But your return came from lowered costs, and there's no tax on that. Nowhere on the IRS tax forms is there a column for "money saved by being smart." So it is truly tax-free. Did making this investment lower your standard of living? Obviously not, except for the inconvenience of having some workmen install the insulation.

Why don't more people do this? Part of the reason is undoubtedly confusion. Instead of "payback period," which is a new concept for many people, and has the prejudice that any payback period greater than 3 years is a hardship on the consumer, we should refer to conservation measures in terms of the *annual return* on investment. So, for example, if you invest in attic insulation, the companies that install it should be free to advertise that your expected annual return will be 17.8%, tax-free.

As president, one of the most important steps you can take is to educate the public about payback period. You need to let people know that a payback period of 4 years is equivalent to a 25% annual return—that a payback period of 5 years is 20%, and even that a payback period of 10 years is a 10% annual return. No other secure investment yields such a high rate. And when you get the public investing in conservation, you will have made enormous progress in improving US energy security.

Invest for a 209% Annual Return, Tax-Free, with No Risk

In the previous section I described a way to get a 17.8% tax-free annual return on your investment. Now I'll describe a way to get a 209% annual return. Why didn't I describe this one first? I was worried that if I did, you wouldn't find the 17.8% return so interesting.

Here's the scheme: replace incandescent lightbulbs with compact fluorescent lights (CFLs).

Some people think that the light of CFLs isn't as pleasant as that from incandescent bulbs; that was true in the early days, but the better bulbs are now much "warmer" in color. A few years ago when I returned to Notre Dame in Paris, I noticed that the chandeliers, once lit with candles, more recently with incandescent bulbs, are now filled with CFLs. The beauty is undiminished. My home is almost 100% compact fluorescents. These days they're even dimmable.

Let's calculate the rate of return on the investment. CFLs are usually said to be cheaper because they last longer, but I'll ignore that for now. In fact, I'll show that they're cheaper even if you ignore their long lives and replace them as often as you replace the ordinary tungsten filament bulb. I'll put in some typical numbers, but you are encouraged to put in your own; you're likely, for example, to be able to buy cheaper CFLs than I assume.

Cost of a 75-watt incandescent bulb:	30¢
Cost of electricity per day (4 hours, at 10¢ per kilowatt-hour):	3¢
Cost of an equally bright CFL (22 watts):	$4
Cost of electricity per day (4 hours, at 10¢ per kilowatt-hour):	0.88¢
Savings per day:	2.12¢
Initial cost difference:	$3.70
Payback period (cost difference/savings per day)	174 days = 0.48 years

Effective interest rate:
(100% ÷ payback period): 209%

If you add in the short lifetime of a tungsten filament bulb, the value of the CFL is substantially greater. A tungsten bulb lasts typically 1,500 hours; beware of "longer life" bulbs because they produce much less light (check the lumen rating, which gives the visible light intensity). That means that over the 10,000-hour lifetime of the CFL, you would have to buy more than six ordinary bulbs, at a total cost over $2. So the actual price differential is $4 − $2 = $2, not $3.70.

Have you switched to CFLs? If not, why not? Maybe you have bad memories of the cold color of the old fluorescents. Or is it that the savings are too small? The cost of operating an inefficient tungsten bulb is only a little more than 3¢ per day. With each CFL, you save less than $1 per month. So who cares?

People in the developing world care, maybe because they are more sensitive to the savings. In my travels in Morocco, Paraguay, Kenya, Costa Rica, and Rwanda, I was amazed to see how common CFLs were. When $1 per month makes a difference, people notice.

For the home owner, 10 CFLs add up to $10 per month, $120 per year. For big companies that invest in millions of bulbs, the 209% return is spectacular. And the value for the United States in energy can be large.

There's more technology on the way. In my own home I've begun the switch to LEDs (light-emitting diodes). They are a bit more expensive than CFLs, but they should last 20 years.

Government Energy Productivity Policy

When you are president, in addition to cajoling your public into making 17.8% or 209% safe, tax-free investments, and educating them about the payback formula, there are some other things you can do. The smartest is one that has already been implemented in California and is being tried or proposed in over a dozen other states.

It's called, somewhat obscurely, *decoupling plus*. The idea is to get the utilities to make the investment on behalf of the public, and to share in the enormous profits.

The best way to explain this is through an overly simplified example. Suppose we live in a hypothetical state that produces and uses 30 gigawatts of electricity. This state will be similar to California, but to make things simpler I will assume that the state knows its future needs precisely and can build new power plants very quickly. I will also assume that the money to do this is available without loans or bonds—that the utility just has extra cash available. This is a typical physicist simplification, but it has the advantage of showing the essence of the issue without the distracting complications of real-life financing.

The state is growing, and next year it will require 31 gigawatts—so the utilities plan to build a new 1-gigawatt plant. Let's assume it's a nuclear plant. For such a plant, virtually all the cost is in building it; the operating costs are low; for our purposes, we'll assume they're zero. Let's say that the plant costs $10 billion to build. If we assume that the cost of electricity is 15¢ per kilowatt-hour (that's correct for California), then yearly revenues will be about $1.3 billion.[9] So the return on the utilities' investment will be 13% per year. That's pretty good. Of course, in the real world most of that 13% would not go to the utilities, since they don't usually have cash, but to the bondholders who financed the plant.

Now the state steps in with an alternative proposal. Because conservation is so effective, the state suggests to the utility that, rather than building the new plant, instead it should invest money in conservation measures. For example, the utility could subsidize energy-efficient refrigerators, better air conditioners, compact fluorescent lights, or the installation of insulation. As we explained earlier, such investments have huge returns, but those returns usually go to the consumer, not to the utility. The most accurate estimate, based on the California experience, is this: the return is about 2.5 times greater than the return on a new power plant. That is a very important rule of thumb. You will use it often when convincing utilities that it is in their best interest to do this.

So instead of investing $1 billion, the utility invests $1 billion/2.5 = $400 million—for example, in subsidizing energy-efficient refrigerators. That reduces the need next year from 31 gigawatts to 30, so no new power plant needs to be built.

Why should the utilities do this? Because if they do it and actually improve the efficiency of energy use in the state (the rules for certification in California are quite strict), the state promises, in return, to raise the electricity rates from 15¢ per kilowatt-hour to 15.05¢. That means that the revenues of the utilities will go up from $30 billion per year to $30.1 billion. In effect, it is a $100 million return on a $400 million investment—an effective interest rate of 100/400 = 25%. That beats the 13% return the utilities would have gotten from building a new plant, so they agree to take this approach.

But isn't the public cheated? Their rates have gone up! Yes, the rates have gone up, but the *bills* from the utilities have gone down, at least on average from what they otherwise would have been. Their electricity use, instead of being 31 gigawatts, is only 30 gigawatts—a 3.3% drop from what otherwise would have happened. So the average use (through conservation) reduces consumption by 3%, and the rate of payment has gone up from 15¢ per kilowatt-hour to only 15.05¢; that's a rise of only 0.3%. The amount paid by the public has actually decreased (from what they would have paid) by 2.7%!

This seems impossible. The public's utility costs have gone down, and yet the utilities are getting a much higher return on their investment than they otherwise would have gotten. Win-win? It is not impossible. It is simply a way to take advantage of the huge returns available on investment in conservation. Because the public is not yet educated enough to make the investment on their own, we induce the utilities to make the investment, and they get to share in the profit.

Invented and named by Art Rosenfeld (physics professor and former California energy commissioner), this is the approach known as decoupling plus. The term *decoupling* comes from the fact that the utility profits are no longer coupled to the construction of new power plants. *Plus* refers to the fact that the rates go up when the utilities successfully invest in conservation. The method has been

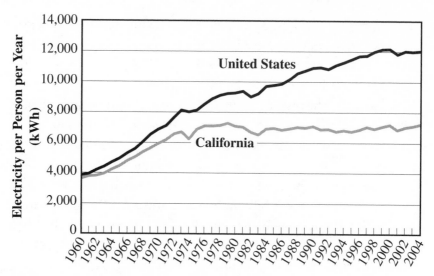

Figure II.13. Yearly per capita sales of electricity, in kilowatt-hours per person.

so successful that Rosenfeld has received many top awards—for this and his other conservation work—including the Enrico Fermi Prize (the highest scientific award given by the US government), and the Global Energy Prize from Russia (worth about a half million dollars—all of which Rosenfeld donated to further research in energy productivity).

For a sense of how successful decoupling plus has been in California, look at the chart in Figure II.13. The electricity used per person in California has not grown since 1980, while the energy use per person of the entire United States has grown by 50%.

There is a catch. If having more efficient lightbulbs means that people use more of them and make their homes brighter—then the decoupling plus program doesn't decrease energy use, and the additional gigawatt is still needed. That's why part of the program includes a careful evaluation and monitoring program. The profit to the utilities is not based on efficiency alone, but on lessened power requirements.

Other Great Investments

I've discussed insulation and compact fluorescent bulbs in detail. Let me now give some additional examples that show the wide range of possible investments in energy productivity.

COOL ROOFS

Look at the roof shown in Figure II.14. In color, you would see that the tiles are somewhere between orange and brown, similar to terracotta. You might never guess that this roof reflects more than half of the sunlight—half of the incoming heat—that hits it. That's because more than half of sunlight is infrared heat radiation,[10] invisible to the eye but not to your skin, and not to the roofs of houses. If you make the surface just right, you can reflect the heat without affecting the visible color.

Are you thinking of installing solar cells on your roof? If you use air-conditioning in your home, you are likely to get a higher return on your investment if you put up a cool roof instead. You will

Figure II.14. A "cool roof." A thermoplastic coating reflects heat radiation.

reduce the heat absorbed on your roof by a factor of 2, and you can save considerable money on the energy you didn't need for the air conditioner.

Even better are white roofs, but many people consider them too bright. When you are president, you may want to encourage white roofs on all buildings whose roofs are not visible from the ground—for example, commercial buildings with flat roofs. You can start with government buildings, but you could also implement an energy productivity policy with the utilities similar to the decoupling plus scheme.

More Efficient Autos

Improvements in automobile efficiency can also be a great investment. When I was young, typical auto mileage was 16 miles per gallon. Now the US average for autos is 30 miles per gallon. What's the difference? At 10,000 miles per year, the difference in the amount of gas you'll need is 292 gallons per year, and at $3.50 per gallon, that's $1,020 per year. If you paid $10,000 extra for that efficiency, then your payback period would be 10 years, giving an effective interest rate of 100% ÷ 10 = 10% per year.

The public thinks that automobile efficiency leads to poor performance and unsafe cars. That can be true—you can certainly get higher miles per gallon by making tiny cars that are both unsafe and uncomfortable; such cars are abundant in Europe (where autos average over 50 miles per gallon). But there's another way. Automobile engines are terribly inefficient when rapidly accelerating; they get only a few miles per gallon. By clever application of the hybrid technology, you can get high efficiency during acceleration by using a battery booster. That's one of the biggest advantages of the hybrid approach, and the source of much of the fuel savings. That's why hybrids often show more miles per gallon for city driving than they do for highway—because city driving, with its stop-and-go nature, has a great deal of otherwise inefficient acceleration.

Although a hybrid engine can give you a good return on your investment, beware. The same is not true for plug-in hybrids or all-electric autos. We'll discuss this further in Chapter 16. When you include the finite lifetime of the batteries (or, equivalently, their depreciation), the true cost of all electric vehicles soars.

Another way to get high performance at low gas consumption is to make the autos of lightweight materials. This approach has been widely opposed by the public because of the widespread belief that heavier cars are safer, at least for the people who ride in them (as opposed to the people they crash into). It is true that making a car lighter by using thinner metal for the body can result in greater danger for the passengers and driver in a crash.

But there are other ways to achieve safety. Scientists Tom Wenzel and Mark Ross at the Lawrence Berkeley National Laboratory have studied the issue in detail. They did find that heavier cars are safer, but that there are far more important features to look for, particularly quality of design and manufacture. For example, the heaviest cars of the conventional Big Three US brands (Ford, Chrysler, General Motors) were safer than their own lighter cars, but remarkably they were no safer than the lightest of the Japanese or German cars. Wenzel and Ross reached this conclusion after a great deal of careful work, including making sure that the comparison took into account drivers' ages, locations, and other systematic differences.

How can you tell whether an auto you're considering buying is safe? Wenzel and Ross uncovered an amazing method. Find the estimated resale price of the car after 5 years. Of course, you can't do that for a new car, but you can probably guess it by looking at the resale prices of similar models that are now 5 years old. The chart in Figure II.15 shows their astonishing finding: the strongest correlation was not between safety and the original price, but between safety and resale price. Original price often indicates luxury, but resale price indicates quality—and, it turns out, safety.

The important conclusion is that lightweight cars can be very safe. What can you do when you are president? Encourage the use of ultrastrong lightweight materials such as thermoplastic composites, carbon, and nylon fibers. These materials could be made much more competitive, perhaps by mandating that cars achieve higher miles per gallon, but doing so with careful tests of safety so that the public

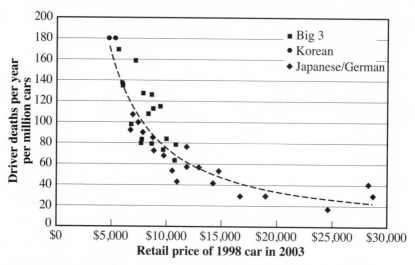

Figure II.15. Cars with high resale value are safer, regardless of price when new.

is aware of the facts. Many people have the misimpression that metal is both stronger and safer than plastic. That prejudice keeps people from buying "plastic" cameras, for example. Your job as president will include educating the public on this subject, and making sure that tests are done.

ENERGY-EFFICIENT REFRIGERATORS

Refrigerators sound like a dull topic, but they give a wonderful example of energy efficiency leading to win-win solutions. In 1974 the average refrigerator in the United States had a volume of 14 cubic feet; in 2012 the average is 23 cubic feet. Do these higher-capacity refrigerators use more energy? No! The newer refrigerators use 72% *less* energy. Electricity costs for the consumer dropped by $180 per year. Moreover, the purchase price dropped by 50% (measured in constant dollars). What drove the change? Government-mandated efficiency, combined with market competition. The new refrigerators have better insulation and use more efficient motors, reducing electricity waste. A similar change is now being implemented for air conditioners (which are really just refrigerators for rooms).

This change had a huge effect on US energy needs. If today's refrigerators had only the 1974 level of efficiency, the United States would need an additional 23 gigawatts of power plants to provide the current level of electric power.

THE MCKINSEY CHART

Are there other examples of potential improvements in energy productivity that are profitable? Lots. The credibility of the field was given an enormous boost when the esteemed consulting firm of McKinsey & Company got into the business in a serious way. The analysis could help countries reach compliance with the expected treaties to reduce carbon dioxide emissions—what was called "carbon abatement." The firm carefully studied the cost of reducing carbon emissions and issued a series of reports. The most famous result was the chart shown in Figure II.16.

The McKinsey chart has two parts. On the left, where the bars go downward, are the actions that are *profitable*—approaches that will make you money while reducing carbon emissions and energy use.

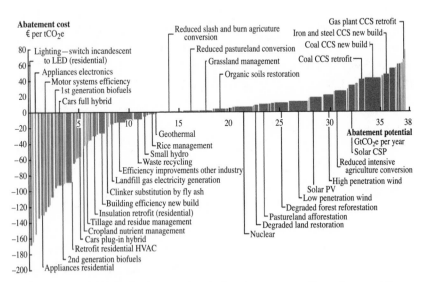

Figure II.16. The McKinsey chart shows that many approaches to reducing carbon emissions can be profitable.

These are things that give a return on investment (McKinsey estimates a 4% annual return), even if you don't care about atmospheric carbon dioxide. They include some of the things I already discussed, such as improved insulation in existing homes and buildings (the chart calls it "insulation retrofit") and more efficient lights (compact fluorescents versus conventional tungsten-filament lightbulbs). Other items on the list include improved efficiency of home appliances, and more efficient HVAC (heating, ventilation, and air-conditioning) systems. It is a complicated chart, but the message is that a lot of money can be saved by conservation and improved efficiency.

On the right-hand side, with the upward bars, the McKinsey chart shows the ways to reduce carbon dioxide that will cost money. These include energy generation methods that are more expensive than coal but don't emit carbon, such as nuclear, solar, and wind. They also include *carbon capture and sequestration*—CCS for short. Sequestration is the process of putting the captured carbon dioxide underground in locations where it will remain for thousands to millions of years. Some people take CCS to stand for carbon capture and *storage*. This means the same thing, and you'll see both.

The chart has one major disadvantage: it is cluttered and confusing. Nevertheless, I include it in this book because of its importance and the far-reaching influence it has already had in making people aware that conservation can be profitable. The profitable side (on the left) makes almost enough money to fund the expensive side (on the right). This means that we actually could reduce carbon emissions without net cost.

There is one point on McKinsey's curve that I disagree with: the claim that plug-in hybrids can save money. I'll discuss plug-in hybrids at length in Chapter 16. Batteries have a finite lifetime; when you include the cost of replacement, plug-in hybrids are not profitable.

Amory Lovins, in his book *Natural Capitalism* (coauthored with Paul Hawken and L. Hunter Lovins), gives an illustrative example showing how much hidden energy productivity potential there is. He presents the following fascinating case history:

In 1981, Dow Chemical's 2,400-worker Louisiana division started prospecting for overlooked savings. Engineer Ken Nelson set up a shop-floor-level contest for energy-saving proposals, which had to provide at least a 50 percent annual return on investment (ROI). The first year's 27 projects averaged 173 percent ROI. Nelson was startled, and supposed this bounty must be a fluke. The following year, however, 32 projects averaged 340 percent ROI. Twelve years and almost 900 implemented projects later, the workers had averaged (in the 575 projects subjected to audit) 204 percent ROI. In later years, the returns and the savings were both getting *larger*—in the last three years, the average payback fell from six months to only four months—because the engineers were learning faster than they were exhausting the cheapest opportunities. By 1993, the whole suite of projects taken together was paying Dow's shareholders $110 million every year.[11]

By now, you probably aren't surprised. Efficiency saves money. In his book, Lovins gives examples of motors, valves, pumps, rooftop chillers, and much more. A motor that is 95.8% efficient was found to be cheaper than one that was 91.7% efficient. It is easy to find more efficient devices, but only if you look. Dow Chemical previously had not.

Feel-Good Measures That Don't Necessarily Work

In implementing energy policy, beware—some measures you can take sound good but, upon more careful consideration, don't accomplish much. In the worst case, they can actually be counterproductive. Many expensive conservation schemes are pushed because they make people feel good. It is important to identify the popular measures that really don't accomplish any good. Some measures can save energy but only in limited circumstances. A surprising example is public transportation.

BUSES

Many people take it for granted that increased public transportation is the obvious solution to our overreliance on autos. Buses carry more people per pound than does a car, so you might think they would get more miles per gallon per person. And they would—if buses were always full. Public transportation can save enormous energy when used in crowded urban environments, but ironically, it can waste energy when used in suburbs and countryside.

If buses don't fill up, and if they have to make round-trips during rush hour—empty on return—then they might not save net energy. You can improve efficiency by using smaller buses during the slow hours, but then you run the risk of having to pass by people at some stops. It's hard to win. Moreover, buses make lots of stops and starts—and that frequent acceleration is when fuel efficiency drops enormously. If the bus has to take circuitous routes in order to get close to where passengers live, the passengers wind up traveling greater distances than if they drove their own cars. To get people into the habit of riding buses, buses must be available all during the day without long waits; otherwise people will take their cars. All this means that a lot of the buses spend a lot of time with only a few passengers.

A detailed study of bus transportation in suburbs around the country done by the Institute of Transportation Studies at Berkeley found that the average break-even point is determined by population density. If there are more than 15 households per acre, then public transportation works. If the population density is lower, then using buses increases energy use. This means that if you live in a suburb with quarter-acre or even one-tenth-acre lots, your community will not save energy by establishing bus routes.

So beware of simplistic solutions such as "more public transportation." There are often hidden subtleties and unintended consequences.

RECYCLING PAPER

There are a lot of fads and useless measures. Some of them may have psychological value—such as schoolteachers asking children to recycle paper in order to increase their awareness of waste. In fact, virtually all paper used in the United States comes from trees specifically grown for that purpose, so recycling paper doesn't save trees. It is also biodegradable, although not readily so when used as landfill. Of course, when used as landfill, it is sequestering carbon taken from the air by the trees grown to make it. Not that it matters much, but recycling paper—unlike burying it—doesn't sequester carbon dioxide.

So, recycling paper neither saves trees nor reduces greenhouse emissions. There's nothing wrong with recycling paper. The key issue is how you justify it. Recognize that people who are misinformed about its virtues are likely to be upset when they find out that they've been misled. People don't like to be fooled, and when they learn the truth they sometimes react badly.

Power Blackouts

It is remarkable how a power outage can disrupt our lives. Whether it's just a fuse blowing, a local transformer burning out, or a huge regional blackout, loss of electric power feels like a return to the Stone Age, or at least to the 1800s. For a really large blackout— affecting your neighborhood or the entire Northeast, for example— it's not just the TV program you miss. There are serious threats to our hospitals, emergency services, communications, and safety. Without streetlights the outdoor world suddenly seems dangerous, and it is.

Look at the two satellite images in Figure II.17. The first one, taken on August 13, 2003, shows the bright lights of the New York City–Boston region. The second one shows the wide reach of the blackout that took place two days later.[12] An event like this

Figure II.17. The northeastern United States lit normally (left) and under blackout conditions (right). The images were taken two days apart in 2003.

makes many people realize that civilization is more fragile than they thought. How could a blackout of this magnitude happen?

The root cause is the *grid*—the system that links together large numbers of power plants, transmission lines, transformers, and users. In normal operations this interconnection provides reliability; if your local plant has a problem, you don't lose power, because other plants readily make up the difference. These interconnections are largely responsible for the fact that electric power in the United States is so reliable. Travel in the developing world and you discover that outages are frequent, in many places daily.

The problem is that the grid is run as a demand-driven system. You don't have to place an order to buy electricity; you just take it when you need it. And unlike water, which can also run out, electricity doesn't fail gracefully. It wouldn't be so traumatic if your lights just dimmed a little or if your motors just slowed a bit, but that doesn't happen. Instead we get widespread failure and complete shutdown.

Power plants are not like water tanks. They don't just hold energy; they make it as you demand it. Operators use historical and current data and weather information to try to anticipate the demand. Natural-gas and hydropower plants can respond to increased demand almost instantly, coal plants are slower, and nuclear is the slowest of all. The problem arises on days when the system is at maximum capacity. If one plant fails, then the others are suddenly required to deliver more power. It could be that one additional air conditioner—the straw that breaks the camel's back—or it might be a needed shutdown of one plant on the grid because of a glitch.

Suppose you hook up too many appliances in your home. The power lines outside deliver constant voltage, and when you turn on the air conditioner, they deliver more electric current. There is a danger that high current will overheat the wires in your walls, so your home has fuses or circuit breakers to cut off the power before they overheat. This seems stupid. Why not just have the system limit you by telling you not to plug in that electric toaster when your home circuit is near capacity? Or maybe just turn off the refrigerator for a few moments—it will stay cold enough—while the toaster toasts? That would be smart.

Our grandchildren will probably take such intelligent house circuitry for granted, but we don't have it yet; we are still in the electrical Stone Age. So is our power grid. When too many people turn on their air conditioners, the draw on electric current from the generators gets too high. There is a serious danger that the generators will overheat, and the operators (or the automated systems that monitor the load) shut them down. It's stupid, but that's the way we do it.

The problem is that if one plant fails, there is a sudden load on the others. If they are already at maximum capacity, then they have to shut down too. The collapse spreads like a chain reaction. The industry calls it a "cascading blackout." That's what happened in the New York region in August 2003.

It doesn't have to be that way. One solution is called "load shedding." When the stations are near capacity, the utilities take turns denying power to local regions one at a time, with the goal of making sure the power plants never actually reach the tipping point. In California this approach was used in 2000–01 and was called "rotating brownouts." Regional customers were given a schedule of when they might lose power so that they could prepare. It worked well but was hated. Power stayed on for emergency services, but businesses and retail stores were basically shut down.

Another solution, suggested by Bill Wattenburg, is to have the utilities decrease the voltage on the line. Wattenburg did a series of tests and discovered (as he had suspected) that most of our home appliances do degrade gradually; air conditioners continue to work at lower voltage, just at a reduced level. They can operate at these lower levels indefinitely without damage.

Another solution is to build more power plants, enough to cover the requirements on the peak demand days. This is a very expensive solution, but it works; it's what was done in California to avoid the rotating brownouts. Small power stations, typically natural gas because of their quick response time, are added to the system to cover the hot days.

That's a very poor investment, largely because the extra power plant is used for only a few days per year. Here's an example that shows why. Assume you can build a 100-megawatt "peaking" station for $100 million. The sole purpose of such a facility is to prevent power blackouts by providing intermittent power only when the grid is stressed. Assume that the station delivers 100 megawatts for 10 days per year, during only the hottest 5 hours each day. That means it operates for only 50 hours each year. The total power it delivers is 5 gigawatt-hours; and at 15¢ per kilowatt-hour, the investors will get only $750,000 per year on their $100 million investment. They are paid back at an interest rate of 0.75%. That's a terrible investment—but good for the politicians if it prevents voter anger.

There must be a smarter way to do this, and there is. It's called the *smart grid*.

The Smart Grid

My favorite implementation of the smart grid is based on market forces. Instead of keeping the cost of power constant, raise the price when the demand is high. This is called *dynamic pricing*. Charge more per kilowatt-hour on hot days; reduce the rates in the evening. When the power is near its peak, the rates could surge from 10¢

per kilowatt-hour to whatever is needed to reduce demand, maybe even $10 per kilowatt-hour. People who really need the power—shopkeepers, manufacturers, and so on, for whom a stop and restart would cause severe difficulties—could keep on going but would have to pay the price.

A problem with this approach is that the price might have to change very rapidly, from hour to hour, and most people don't (or can't) monitor the rates, certainly not during the day when they may be at work. The proposed solution is to introduce "smart meters." With a smart meter you could preprogram your home or building to turn down your usage when the price gets too high. You might, for example, decide to turn off your air conditioners if the price rises above $1 per kilowatt-hour. Or you could set your system to turn off only those appliances that you can temporarily do without; for example, you might set your air conditioner and electric clothes dryer to turn off but leave your lights and burglar alarm on.

In many homes, power enters at two different voltages: 120 for lighting and small appliances, and 240 for air conditioners, washers, dryers, and other devices that pull high loads. A simple way to program the smart meter would be to turn off the 240-volt circuits when the dynamic pricing surges, while leaving the 120-volt circuits alone.

A CASE STUDY: SMART METERS IN CALIFORNIA

If smart meters were so smart, you would think they wouldn't need any marketing to the public. At no cost to the consumer, the utilities would install new electricity meters in your home, and they would have advantages for everyone. They would keep your electric bills low during periods of dynamic pricing. And by using modern electronics, they could be far more accurate than the old meters.

Starting in 2006, smart meters were introduced in California. One goal was to provide information on electricity use for the utilities—just who was using how much power at which times. The public reacted with outrage. There were three strong objections: overcharging, loss of privacy, and danger from microwaves.

The old meters had not been very accurate—possibly half of them overcharging for electricity, and half undercharging. So with half the customers, the rates went down; no complaint from them. Of course, the half who had been previously undercharged saw their bills go up—some by quite a bit. For these unlucky people (or, more accurately, previously lucky people whose luck had run out), the electricity cost rose. The result was a huge number of complaints from people who assumed, incorrectly, that it was the *new* meters that were wrong.

Another problem was the fact that the meters were designed to reduce power automatically in case of an extreme emergency. Newspapers expressed outrage at this public intrusion into the private home. Somehow they missed the point that the alternative was a blackout or brownout—something that is, in a very real way, even more intrusive.

Finally, people attacked the fact that these meters communicated to the utilities by using microwave *radiation*, a word that frightens much of the public. There had been a lot of "controversy" over the dangers of microwaves in cell phones, with many people believing that they could cause cancer, and now the smart meters were yet another source of microwaves, right into the home. Microwaves penetrate walls and wood—and even bricks. It was yet another attack on privacy.

Of course, microwaves have been around for a long time. They are the means for TV broadcast, especially for the broadcast channels above channel 13. They are what heat the food in microwave ovens, and they are signal carriers for cell phones. They are used for computer Wi-Fi connections, and to link your portable home telephone handset to your landline.

How big is the danger? There is a standard way to evaluate it using the methods that are used for nuclear radiation, such as that from Fukushima. The unit is the rem (or the sievert, equal to 100 rem), which we discussed in Chapter 1. Rem measures damage done when radiation passes through the body. For example, it takes a billion 1-MeV gamma rays[13] over every square centimeter of your body to do 1 rem of damage to your flesh. It takes 2,500 rem to trigger a cancer. How many rem does a microwave cause? The physics calculation is well known: zero. Microwaves do not cause cancer.

But what if physics is wrong? What if the best judgments of those who study and evaluate this issue are wrong? What if the dissident scientists who say there is really a bigger threat are right? It's always possible. My concern is that there are so many *known* threats, real threats, in this world, that the wisest course of action is to pay attention to *them* and stop worrying about speculative ones. The biggest dangers in the life of the typical citizen come from smoking, from being overweight, from poor diet, from accidents in the bathroom, from auto accidents, war and pestilence, and so on. Even if the people who are concerned about microwaves are right, these other dangers are many thousands of times more important. If we worry about things that are estimated by science to be negligible, in the thought that science is sometimes wrong, then the number of things to worry about skyrockets, and our actions go completely out of balance.

III

ALTERNATIVE
ENERGY

WE HAVE *been through several energy crises in the last few decades, related primarily to transportation energy. Now we're worried about a new one: the climate change that could be induced by continued fossil fuel use. What can we do? How can we stop? What new sources of energy can we bring forth?*

As with other rapidly developing fields, there is an equally rapidly developing jargon. You've probably heard about green energy, clean energy, renewable energy, *and* sustainable energy. *Nuclear power advocates include it as "green" and "clean" because it produces no carbon dioxide; nuclear opponents argue it is not clean because it produces radioactive waste, and it is not sustainable because the available fuel will run out someday. I'd rather not get into debates over names; let's just look at all of these sources and discuss their advantages and disadvantages. I titled this section "Alternative Energy" to try to encompass all of them.*

THE ALTERNATIVE-ENERGY field is wide, technically complex, and full of uncertainties. When looking at the possibilities, many people find the numbers difficult to deal with, and they base their judgments on their feelings instead. Disagree with people and they might respond with outrage or condescension at your presumed ignorance. Energy politics is becoming dominated by dogma, and it shows signs of turning into a religion, so beware. The field of alternative energy is dangerous territory for any future president to tread. But you must not only learn the field; you need to understand it well enough to lead.

Be on the lookout for *skepticism bias*, exemplified by people who are exceedingly pessimistic about the potential for energy sources that they don't like, but extraordinarily optimistic about their preferred solution. There are enthusiasts for every type of alternative—solar power, wind, nuclear—even though each has its technical difficulties. Hey, we are inventive Americans and we can do it! But express optimism for the wrong thing (wrong in someone else's mind), and you need to be ready for the complaint that it "isn't proven." As president, you will need to evaluate the alternative numerically and make decisions based on the best data available.

You also have to watch out for *optimism bias*. "Yes we can" can be an inspiring mantra, but it doesn't always apply to technology. The rapid advance of computers is often cited as a model—but some other technologies have never moved. Look back at old issues of science and technology magazines, like the one pictured in Figure III.1, and you'll find that flying automobiles for the average citizen have been "just around the corner"—for over 70 years.

Remember that there are two driving issues for energy: energy security and climate change. Some alternative energies address one or the other; some can help with both. Try not to get them confused. In the past, liberals tended to emphasize the climate change aspects, and conservatives the energy security side. But that can change.

It is instructive to begin with a look at the present-day costs of our different energy sources, and how the alternative sources compare. Table III.1 is based on numbers published by the US Energy Infor-

GIRO-CAR
SEE PAGE 87

Grappling With Death Under the Sea
Debunking Poison Gas War Scares

Figure III.1. The cover of Modern Mechanix *for July 1935. The flying automobile reappears on covers of popular science and mechanics magazines every decade or so, illustrating that past failures in prediction do not dampen optimism bias.*

mation Administration. You'll see lots of numbers bandied about, but this is the source that I think does the most honest job; it avoids the hyperbole and bias that is frequently found in the analysis of advocates.

In Table III.1, the estimated cost of producing electricity for each technology assumes that the power plants will be constructed and ready for operation by 2016. The table has much of the information you need to evaluate the economics of the various options. The numbers in this chart are realistic, and I'll use them in the rest of this section to compare different energy technologies. Two caveats: First, the chart assumes that the cost of capital is 7.4% per year; as I write this (in 2012) the interest is much lower, and that reduces the cost for capital-intensive technologies such as nuclear. Second, the chart assumes a carbon emission trading cost for coal and natural gas of about $15 per ton; no such price on emitted carbon is currently in force in the US.

Table III.1. A breakdown of the cost to produce a kilowatt-hour of electricity, for different kinds of power plants.

Plant type	Capacity factor	Capital cost per kWh	Operation and maintenance per kWh	Fuel per kWh	Transmission investment per kWh	Total cost per kWh	Lowest possible cost per kWh
Conventional coal	85%	6.5¢	0.4¢	2.4¢	0.1¢	9.5¢	8.5¢
Advanced coal	85%	7.5¢	0.8¢	2.6¢	0.1¢	10.9¢	10.1¢
Advanced coal with CCS[a]	85%	9.3¢	0.9¢	3.3¢	0.1¢	13.6¢	12.6¢
Natural gas							
Conventional combined cycle	87%	1.8¢	0.2¢	4.6¢	0.1¢	6.6¢	6.0¢
Advanced combined cycle	87%	1.8¢	0.2¢	4.2¢	0.1¢	6.3¢	5.7¢
Advanced CC[b] with CCS[a]	87%	3.5¢	0.4¢	5.0¢	0.1¢	8.9¢	8.1¢
Conventional combustion turbine	30%	4.6¢	0.4¢	7.2¢	0.4¢	12.4¢	9.9¢
Advanced combustion turbine	30%	3.2¢	0.6¢	6.3¢	0.4¢	10.3¢	8.7¢
Advanced nuclear	90%	9.0¢	1.1¢	1.2¢	0.1¢	11.4¢	11.0¢
Wind	34%	8.4¢	1.0¢	0.0¢	0.4¢	9.7¢	8.1¢
Wind – offshore	40%	20.9¢	2.8¢	0.0¢	0.6¢	24.3¢	18.7¢
Solar PV	22%	19.5¢	1.2¢	0.0¢	0.4¢	21.1¢	15.9¢
Solar thermal	31%	25.9¢	4.7¢	0.0¢	0.6¢	31.2¢	19.2¢
Geothermal	90%	7.9¢	1.2¢	1.0¢	0.1¢	10.2¢	9.2¢
Biomass	83%	5.5¢	1.4¢	4.2¢	0.1¢	11.2¢	10.0¢
Hydro	51%	7.5¢	0.4¢	0.6¢	0.2¢	0.9¢	0.6¢

[a]CCS stands for "carbon capture and sequestration" or "carbon capture and storage."

[b]CC stands for "combined cycle," a method that uses both gas and steam turbines.

Source: Based on the Energy Information Administration's *Annual Energy Outlook 2011* (December 2010), http://www.eia.gov/oiaf/aeo/pdf/2016levelized_costs_aeo2011.pdf.

The most interesting columns are the last two, which show the cost to produce one kilowatt-hour of electricity by a new plant. The last column shows the cost if the plant is built in an optimum location (for example, a wind turbine in a windy area), and the preceding column shows the cost if the plant is built at a random location (think of that as near a population center). I'll return to these numbers many times in the coming pages, but I suggest you take a quick look them right now. Note that natural gas is the cheapest source of new energy, as low as 6.3¢ per kilowatt-hour (in the next-to-last column). Coal (9.5¢) is cheaper than nuclear (11.4¢), but maybe not as much cheaper as you might have thought; nuclear is expensive to build (the capital cost is high) but cheap to fuel and run. The cost of both coal and nuclear is lower when interest rates are low. Wind is cheap (10¢), unless it is offshore (24¢). Solar is still expensive (21¢–31¢), in part because of its low capacity factor, a consequence of reduced sunlight on cloudy days and none whatsoever at night.

In making the table, the Energy Information Administration assumed a cost for natural gas of about $4.50 per million cubic feet, but in early 2012 the price had dropped to $2.50. If that price holds, the total cost per kilowatt-hour for natural gas would drop from 6.3¢ to 4.3¢. Compare that to the cost of coal: 8.5¢ per kilowatt-hour for new plants at the least expensive locations. Cheap coal no longer looks so cheap.

Besides being cheap and abundant, natural gas has another huge advantage: for equal energy, it produces half of the carbon dioxide of coal. The reason has to do with its chemistry: natural gas consists primarily of methane, with chemical formula CH_4. When this burns, half of the energy comes from the combustion of the carbon C to make CO_2, but an equal amount of energy comes from the burning of the four hydrogens to make water, $2H_2O$. Thus half the energy is produced without CO_2. Right now the developing world depends heavily on coal; to the extent it could switch to natural gas, its greenhouse emissions will be cut in half. In Section V, "Advice for Future Presidents," I'll suggest a specific measure to help accomplish this based on a sharing of shale gas technology with the developing world.

8

SOLAR SURGE

THE PRICE of solar cells is plummeting, and as a result, interest in solar is surging. I predict that in a decade or so, the cost of solar cells will be virtually *negligible*; that is, it will not be a consideration when building a solar power plant, or when installing solar power on your roof. That doesn't mean the total cost of solar power will be negligible; you will still have to pay for installation and maintenance. And you will still need to have a backup for rainy days.

The Physics of Sunlight

Sunlight delivers about a kilowatt of power per square meter onto the surface of the Earth. That's not hard to remember; think of it as ten 100-watt bulbs.[1] Could we use solar power to drive a car? If we had 2 square meters of solar cells on the auto roof and the sun was directly overhead, 2 kilowatts would be incident. The best solar cells convert only 42% of that energy to electricity, so the usable power would be 840 watts. That's 1.1 horsepower, enough to compete with a real horse, but not to meet the needs of most consumers; typical

US autos cruise the freeway using 10–20 horsepower, and deliver 40–150 when needed for acceleration.

On the other hand, with a large area, sunlight can add up. Over a square mile the sun delivers 2.6 gigawatts of power. Convert that at 42% efficiency, and you get over a gigawatt of electricity—the same as the power from a large coal or nuclear plant. On average, however, you don't do that well. Solar plants deliver less power when the sun is oblique, and none at all at night. Even in a cloudless desert, the average solar power is only 25% of the peak,[2] about 250 watts per square meter. Table III.1 uses a 21.7% capacity factor; that presumably includes reduced power from dust on the cells and downtime for maintenance. On the other hand, if you need the power primarily in the afternoon, when both air conditioners and factories are running, then solar can be an excellent supplement.

The trick is to gather all this energy cheaply. One way is to focus it in the same way that you can use a magnifying glass to start a fire. The idea goes back to Archimedes, who, according to legend, used mirrors with sunlight to attack a Roman ship during the siege of Syracuse. For a solar power plant, the high temperature produced by this focusing of the sunlight is used to boil water, which in turn runs a turbine. This approach is called *solar thermal*.

Solar Thermal

My family uses solar thermal when we go backpacking. On a layover day, we leave a plastic bag of water out in the sun. Sunlight passes through the transparent cover and heats the water; by the middle of the day the water is hot, suitable for a shower, as Figure III.2 illustrates. Similarly, in sunny climates transparent pipes installed on a roof can preheat water before it goes to a hot-water heater. When you use solar power for heat, it is essentially 100% efficient.

On a large scale, however, solar thermal makes less sense. Although many solar-thermal plants are being built around the world, to be affordable they depend on subsidies. Their cost is largely

Figure III.2. The author's family enjoying a solar shower on a backpacking trip near Yosemite, some years ago.

in construction, and that cost is unlikely to come down in the near future. In Table III.1 you can see that, spread over the lifetime of the plant, the capital cost of electricity produced from a solar-thermal plant is 25.9¢ per kilowatt-hour. Compare that to natural-gas combined cycle at 1.8¢ per kilowatt-hour. Will that capital cost come down? I don't believe it can come down very much; the cost is in structure—"bricks and mortar"—which is low-tech and unlikely to drop in price.

The most dramatic-looking solar-thermal plant is the solar tower. Figure III.3 shows a 5-megawatt solar-tower facility in California. Each of the 24,000 mirrors must be continually aimed to reflect sunlight from the moving sun[3] to the top of the tower, where the

Figure III.3. Sierra SunTower in California. Twenty-four thousand mirrors direct sunlight onto the central tower. The heat is used to produce steam that drives a turbine to generate 5 megawatts of electricity.

concentrated light heats salt; this in turn heats steam that is used to run a turbine to produce electricity. To collect the light efficiently, the tower has to be high enough (160 feet tall, as high as a 16-story building) to see all the reflectors without them obscuring each other. The 5 megawatts this plant produces can be described as 0.005 gigawatt—in other words, only 0.5% of the electric power delivered by a large conventional (gas, coal, or nuclear) plant. It is a mechanical system full of a large number of moving parts that require maintenance. Imagine the effort required to run 200 of these plants to produce the energy equivalent to the output of one conventional natural gas plant.

More practical, in my estimation, is a solar trough, such as the one shown in Figure III.4. Long, cylindrical mirrors focus sunlight onto a pipe carrying the liquid. The system uses clever optics to minimize the need to repoint. As the sun changes its angle in the sky, the light continues to be focused on the pipe—just on a different part of the pipe. There are few moving parts, other than the flow of the liquid in the pipes and the possible need to slightly adjust the mirror angle as the seasons change.

Spain has been of the biggest developers of solar thermal; its capacity reached nearly 4 gigawatts by the end of 2010—about 3% of its electric power usage. But solar thermal is successful only when subsidized. In 2008 the Spanish government drastically cut its subsidies for solar power and set a limit for construction of new subsidized plants at 0.5 gigawatt per year. The current financial woes of Spain may very well mean an end to its formerly strong interest in solar thermal.

Subsidies can be subtle; sometimes they are more popular with the public when they aren't called subsidies. In 2006 California, the world's eighth largest economy, passed a law known as AB32 that requires utilities to provide 20% of their power from "renewable" energy by 2020. In 2011, this law was updated to increase the percentage to 33%. In effect, this requirement acts as a subsidy for solar power. Here's how. By law, the utilities are guaranteed to make a

Figure III.4. Solar trough at Kramer Junction.

profit; they are allowed to charge the consumer whatever they need to charge. If they're forced to use more expensive energy, they still make a profit; consumer rates rise to pay for it. These price increases are effectively a tax without being called a tax.

In addition to subsidies, solar-thermal plants require sunny days. Diffuse light, from clouds, cannot be focused by a lens or mirror, so it cannot effectively warm water, even on a very bright day.

Solar thermal does have several important advantages that make it useful in certain cases. If it's sunny but you don't need the electricity right away, you can store the hot salt and use it when needed later. So you can run the plant full bore during the heat of the day and use the energy easily the next morning if that's when your factories need it. Another advantage of solar thermal is its remarkable efficiency: because concentrated sunlight can heat the salt to extremely high temperatures, the energy can be converted into electricity with an efficiency exceeding 50%.[4] The efficiencies for trough plants are lower because the concentration of sunlight is not as high, and the liquid must flow through long pipes that can lose heat.

As I said earlier, because the capital cost is so high (25.9¢ per kilowatt-hour), and the prospects of lowering this expense are so low, I consider it unlikely that solar thermal will become competitive. It still has real value for local heating, including systems designed to preheat water for home water heaters, and for solar showers for backpacking. As a large-scale energy supplier, it is viable only as long as it is heavily subsidized. In the end, that's not sustainable.

More likely to succeed, in my estimation, are solar cells that turn the photons of sunlight directly into electric voltage.

Photovoltaic Cells

Solar cells, also called photovoltaic or PV cells, are thin wafers that absorb sunlight and produce electricity directly. They use the physics discovery known as the *photoelectric effect*—a phenomenon that was first explained by Albert Einstein and that earned him a Nobel

Prize. (No, he did not get it for his theory of relativity.) In the photoelectric effect, an incoming particle of light known as a photon knocks an electron away from the atoms that it is normally associated with, and it lands on a metal electrode. When that electron moves from the electrode onto a wire, it is electricity, and it carries with it some of the energy of the photon of light. The fraction of energy turned into electricity by a reasonably priced cell is only 10%–15%—although it might reach as high as 20% in the near future. It is 42% for expensive solar cells, and that percentage could also go up.

It is interesting to know (although not really essential for a future president) that photovoltaic cells are truly quantum devices. The photoelectric principle was one of the foundations of quantum mechanics; although Einstein is often thought of as disliking quantum physics, he was indeed one of the key founders of this field.

Much recent interest in the solar-cell approach comes from the currently plunging cost of the cells. One way to describe the expense is as "cost per installed watt." If the cell will produce one watt of power at peak sunlight, and it costs $7 to produce, then we say the cost per installed watt is $7. That was the actual cost just a few years ago, but the field is high-tech and very competitive—in part because of the subsidies. In 2011 the cost dropped below $1 per installed watt; that milestone was celebrated throughout the alternative-energy community, and the bottom is not in sight. That's very exciting.

It is also misleading. When we talk about the cost per installed watt for coal, nuclear, or natural gas, we mean average power, delivered around the clock. But the convention in solar is that we talk about the cost per installed *peak* watt. Recall that one watt at peak, even if there are no clouds, is ¼ watt on average when you include the varying angle of the sun and its total absence at night. The energy in sunlight is lower still when it's overcast; clouds reflect the light back to space. At a typical location, intermittent clouds reduce the average solar energy by half, so on average a solar cell delivers only one-eighth of the electric power that it can produce at its peak.[5]

Of course, electric power is often more valuable during the day, when you might be running air conditioners or factory machines. So the true value of solar depends on the time of day it is needed. Still, beware of fantastic claims of low cost per watt.

Let's look in more detail at the cost. Assume the cell costs $1 per watt to buy and install. That's for peak power. The average sunlight is one-eighth that, so on average, the cell produces ⅛ watt. There are about 8,000 hours in a year, so that one cell will deliver 1,000 watt-hours, or about 1 kilowatt-hour of electricity, in a year. That's worth about 10¢ to the consumer. Thus the return on the $1 investment is 10¢ per year, or about a 10% return, assuming no installation or maintenance costs. From this you need to subtract the depreciation. If the cell lasts only 10 years, then you're losing value at 10% per year; in 10 years you have to replace the cell and put in another $1. The net result: you get your money back, but with no profit. If the cell lasts for 20 years, then the effective return is 5% per year.

But we have ignored a major cost—that of the electronics that must be added to the cell to make the produced electricity useful. Photocells deliver their power at only a few volts, but most of our home electric devices, from lightbulbs to refrigerators, are designed for 110- or 220-volt alternating-current electricity. The simplest solution is to attach these cells in series, boosting the voltage to the sum. But then if one cell fails, it brings down the whole array. A better and widely implemented solution is to use an electronic device called an *inverter* that converts the low voltage to standard household values.[6] Add the cost of inverters, the cost of installation and maintenance, and the optional cost of backup batteries, and most home and business rooftop installations would show no profit if the government did not heavily subsidize them. Table III.1 shows that the capital cost alone amounts to 19.5¢ per kilowatt-hour.

How can solar compete? I expect the cost of the inverters will drop as the market for them expands and new technology is developed. Large backup batteries, suitable for large commercial PV plants, are getting cheaper. And finally, installation and maintenance costs can be low if consumers don't include their own labor, or if the plants are used in parts of the world with low wages. Ironically, because of the low cost of labor, solar power may surge in the developing world even while is too expensive for the United States.

REMARKABLY, even within the field of photovoltaics, several very different technologies are in strong competition. The most important of these are silicon crystals, cadmium telluride (CdTe), copper indium gallium selenide (CIGS), amorphous (glass-like) silicon, and multijunction cells.

It's not clear how many of the details need to be known by a future president. Let me give a broad overview anyway, in case you're interested.

SILICON

Silicon crystals were the original solar cells, the ones that were used on the first space missions, and still the ones widely used in the home market. Silicon is cheap; it is a major ingredient in sand (silicon dioxide), but purifying it is a major expense.

Just a few years ago, most people were pessimistic about the future of silicon solar cells, in part because growing large single crystals was expensive, and the cells were not very efficient. In 2007, silicon solar cells cost about $5 per installed watt, but in recent years that price has dropped dramatically. In 2010 the price dropped below $2, and by 2011 it was below $1.

There are two reasons for the rapid plunge in cost. The first is that it was *possible*; that is, technology for cheap solar cells could be found. There is nothing fundamentally expensive about silicon. Not all technologies can be made cheaper. The cost of computer chips has dropped dramatically in the last few decades, but the cost of many other technologies has not; for example, lead-acid batteries have stabilized in price. It turns out that solar cells have followed the computer chip path. The other reason for the drop in cost is that there was *competition*. Thanks in part to "renewable energy" legislation, there has been a huge demand for carbon-free power. Given the existence of a market, investors were willing to take risks on technologies that could compete.

The largest manufacturer of solar cells in the world is now Suntech Power in China; this company produces cells with efficiencies of up to 15.7%. There are constant complaints in the United States that Suntech's cells are being sold below cost in order to drive US manufacturers out of business. Suntech is now producing more than 1 gigawatt (peak) of solar cells every year. That is a stupendous achievement. First Solar in the United States is a close second. But to put 1 gigawatt per year into perspective, recognize that because of nights and cloudy days, the average power output of a year's worth of Chinese solar cells is not 1 gigawatt, but only ⅛ gigawatt. China is installing about 50 gigawatts of coal power every year—400 times greater than the added solar capacity. So although solar sounds big, it is way, way behind. It will take enormous growth for solar to really become a substantial contributor even in the Chinese market, let alone the world market.

CdTe (Cadmium Telluride)

Tellurium has almost no commercial value—except that its compound with cadmium, CdTe, has a superb ability to absorb sunlight and release electrons. A layer only 3 microns thick (about one-tenth the thickness of a human hair) can produce electricity with an efficiency of 15% or more. Moreover, CdTe can be deposited on thin sheets, yielding flexible solar cells that don't have the crystalline fragility of silicon cells (which are typically 30 times thicker).

CdTe is used by First Solar, the largest solar-cell manufacturer in the United States. Already First Solar is producing over 1 gigawatt of solar cells each year, and it is growing rapidly. The company says its price per installed watt in 2012 dropped to 73¢, but it's hard to know since it depends on how they amortize factory construction.

There is a serious concern that we will run out of tellurium. That worry is based on the fact that only about 800 tons per year is produced, mostly as a by-product of copper mining. It takes about 100 tons to make a gigawatt solar plant, so the world's supply will allow only 8 gigawatts per year. First Solar is soon going to reach 2 gigawatts per year, and in a few years, if its growth continues, it may be using the world's entire yearly supply. Some experts believe that the low production of tellurium simply reflects the fact that until recently there was no market for it, and they think we will find abundant new sources of tellurium as the demand for solar cells increases.

Another worry about CdTe is that cadmium is highly toxic. Proponents argue that the cadmium is safely confined in the cell, but there is always danger of release in a fire, particularly if the cells are installed on a roof. Studies of this hazard show that such release is not very likely, but the public will undoubtedly continue to be concerned, particularly if the competitors (CIGS and silicon manufacturers) keep raising this issue publicly.

CIGS (Copper Indium Gallium Selenide)

The names of the four main constituent elements—copper, indium, gallium, selenium—are too ponderous, so people refer to these cells by the acronym CIGS (pronounced "sigz"). Like CdTe, CIGS absorbs sunlight so readily that the cells can be made very thin. One form of manufacture is amazing: small beads are deposited on a metal-coated glass or plastic using a device that looks and behaves like an ink-jet printer. Once in place, the material is "sintered" (treated with heat), causing the pellets to fuse. Finally, additional layers are deposited and treated. In the end, the entire structure is only 3–4 microns thick, just as with the CdTe cells.

CIGS cells have the advantage over CdTe cells in that they don't contain any highly toxic material, but they have the disadvantage that one essential ingredient, indium, is in very high demand and short supply. Indium tin oxide is a transparent conductor of electricity, and for that reason it is used on virtually every modern TV and computer and game display. Some estimates say we will run out of indium in a decade or two, even if we don't use it for solar cells. But optimists argue that there is really a lot of indium available, if the demand grows.

One of the stars of CIGS technology is a company called Nanosolar, with a large factory in San Jose, California, currently producing over 640 megawatts of solar cells each year. Nanosolar's current efficiencies are just over 10% (compared to 15% for silicon) but it has achieved 20% efficiency in the lab. Efficient cells are particularly important for locations that have limited area, such as rooftops. If you have plenty of room to spread things out (for example, in the desert) than the key number is cost per watt.

Nanosolar and other thin-film companies are suffering from the sudden and surprising drop in the cost of Chinese silicon cells—which also happen to have higher efficiency. US imports of Chinese solar cells increased 15-fold between 2006 and 2010, primarily because of the drop in prices. Some people think that China is heavily subsidizing its silicon solar industry in order to grab market share and drive competitors out of business, and politicians in the United States have called for action to protect our industry from such unfair competition.

CIGS got a bad name in 2011 because of the bankruptcy of Solyndra, a CIGS-based company that had received over a half billion dollars in loan guarantees from the US government. Solyndra blamed its bankruptcy on Chinese competition; the company said that the Chinese solar cells were being subsidized (ignoring, in its argument, that it, too, was subsidized). However, a deeper reason may have been the complexity of the Solyndra design. Solyndra put its CIGS cells inside a hollow glass cylinder (hence the name of the company). Although Solyndra's web page claimed this would increase efficiency, it was not hard to show that it actually decreased efficiency. Solyndra also claimed that the cylindrical design would help in windy conditions and make the cells easier to install. When I reviewed Solyndra technology about a year before its bankruptcy, I concluded that the wind resistance had no significant benefit, and what value it did have could be easily matched with small innovations applied to the standard flat-cell geometry. I believe that the true reason Solyndra could not compete with the Chinese competition was not because of subsidies or lower labor costs in China, but because the Solyndra design was inherently more expensive to manufacture and the cylindrical geometry lowered the average light intensity hitting the cells.

MULTIJUNCTION CELLS

If cost doesn't matter, or if space is critical, you can use a multi-junction cell, typically made of gallium arsenide (GaAs), germanium, indium, and other metals and semiconductors. These cells have multiple layers, one for each wavelength range in the solar spectrum, and as a result they reach very high efficiencies—already 42%, and possibly higher in the future. That number is their big selling point. Because they deliver more power per area and per pound than anything else does, they are widely used in space. They are the solar cells used on the Mars rover. But they are very expensive, typically costing $500 for just one square centimeter.

There is a trick, however, that allows the use of multijunction cells at low cost. Buy a small cell, and use lenses and mirrors to concentrate the light onto that area by a factor of 500–1,000; the result can be a good number of watts per dollar. This method, similar to the method used in solar-thermal plants, is called *concentrator PV*. To prevent the cells from overheating, good thermal conductors must be attached to the cells to conduct away the heat.

Multijunction cells have dropped in price to the point that concentrator PV is now a competitive technology, and several firms have started to produce them. The disadvantage is that they have to be carefully pointed at the sun, so they must move as the Earth rotates. If there's a cloud between them and the sun, they don't work at all; the concentrator works only with pure, unscattered sunlight. The manufacture of the pointing system and the reflectors has become the major cost, not the price of the solar cells. The success of this approach depends on reducing these expenses and on achieving low maintenance costs.

The major value of the concentrator PV approach is the high efficiency. If you have limited space (for example, a rooftop), these cells can deliver 2–4 times the number of watts as its competitors.

Concentrator PV is still under development. SolFocus has raised over $170 million; it has two products: a cell array that produces 6.1 kilowatts, and another that delivers 8.4 kilowatts. A company named GreenVolts raised $39 million in late 2011 to build equipment to make a 1-megawatt system. It is also possible to design small systems appropriate for small areas such as rooftops; a start-up called Sun Synchrony uses small modules (just a few inches in size) that aim themselves toward the sun automatically. If we take into account all of the inefficiencies, these modules extract over 30% of the power in sunlight and convert it to electricity.

Solar-Cell Summary

The solar-cell field is intensely competitive and developing fast. Prices are dropping so rapidly that the winners are likely to be decided by criteria other than solar-cell price, including cost of installation, cost of maintenance, cost of conversion to household voltages, lifetime of cells, and efficiency.

The installation and maintenance costs are likely to be lower in a developing country such as China where wages are low. For that reason, I suspect that the truly rapid breakout of solar power will be in such countries. That's good news for people concerned about global warming, since most of the future greenhouse gases will come from developing countries. If we're going to limit such emissions, we need an energy technology that does not produce carbon dioxide and that the developing world can afford. Solar could be that technology. But keep in mind the huge gap between the solar cell production rate in China (1-gigawatt peak of cells manufactured each year—equal to $1/8$-gigawatt average) and the introduction of new coal plants (over 50-gigawatt average per year).

9

WIND

MODERN wind turbines are enormous. Look at the 7-megawatt wind turbine shown in Figure III.5. Next to it is a photo of the Statue of Liberty—to the same scale. The total height of the turbine is 650 feet. The Statue of Liberty is only 305 feet.

When they are seen from a distance, people sometimes don't realize how huge these monsters are. Look at the next image (Figure III.6), showing me standing next to a wind turbine blade that has not yet been installed. Manufacturers of such blades face a serious problem: to make them sturdy they are assembled in the factory, but because of their size, transporting them to the site is extremely difficult.

Why are the turbines so huge? As anyone who has flown a kite knows, the good winds are up high. Winds at an altitude of 200 feet are typically twice the velocity at 20 feet. The power increases as the cube of the wind velocity,[7] so with 2 times the wind you get 2 × 2 × 2 = 8 times the power. If the wind increases by a factor of 3, then it increases the power by 3 × 3 × 3 = 27. That makes a huge difference, so it is worthwhile to build a tall structure to reach those high kite-flying winds. Tall structures are expensive, so you also make the blades big to intercept as much wind as possible. The cube law is

Figure III.5. A large modern wind turbine is much taller than the Statue of Liberty. This 7-megawatt turbine was manufactured by Enercon in Germany.

Figure III.6. The author and a wind turbine blade, in Idaho.

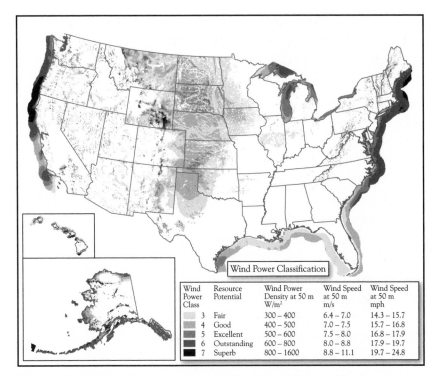

Wind Power Class	Resource Potential	Wind Power Density at 50 m W/m²	Wind Speed at 50 m m/s	Wind Speed at 50 m mph
3	Fair	300 – 400	6.4 – 7.0	14.3 – 15.7
4	Good	400 – 500	7.0 – 7.5	15.7 – 16.8
5	Excellent	500 – 600	7.5 – 8.0	16.8 – 17.9
6	Outstanding	600 – 800	8.0 – 8.8	17.9 – 19.7
7	Superb	800 – 1600	8.8 – 11.1	19.7 – 24.8

Figure III.7. Map of wind resources in the United States. The best locations are off shore; they have strong wind and are near cities, but they are expensive to build. The white areas have average winds too low to compete with other energy sources.

absolutely key to understanding wind power. It shows that wind is an extremely poor source of power in areas (and altitudes) with low velocity, and extremely good where it is windy.

Over the United States, wind velocities average about 5 miles per hour near the ground. That is too slow to be very useful. The map in Figure III.7 shows areas with higher values. All the shaded areas carry economically recoverable wind power. The greatest resources are in the Great Plains and in the coastal waters.

Here's a simple equation that relates the wind velocity (in miles per hour) to the power density (in watts per square meter):

$$\text{Watts per square meter} = \frac{(\text{Miles per hour})^3}{10}$$

So, for example, if the wind speed is 5 miles per hour (mph), then the power density is $(5 \times 5 \times 5)/10 = 12.5$ watts per square meter. Pretty low. If the wind velocity is 18 mph (as is found in much of the Great Plains), then the power density is $(18 \times 18 \times 18)/10 = 583$ watts per square meter. In offshore regions, wind velocities of 24 mph yield 1,382 watts per square meter. The cube law makes a huge difference.

Look again at the German wind turbine shown in Figure III.5. Each blade is 63 meters (206 feet) long. When the blades spin, they sweep out an area of $\pi R^2 = 12{,}462$ square meters. If the wind blows at 20 mph, then our formula says the power density is $(20 \times 20 \times 20)/10 = 800$ watts per square meter. Multiply that by the area to get 10 megawatts.

The physics of blowing wind is complicated, and it can't be computed with pencil and paper; big computer programs are needed to optimize the blade design. Consider the following engineering miracle (at least *I* think it's a miracle): even with just three blades, a modern wind turbine can intercept and suck more than half the energy out of wind that blows through the circular disk defined by the diameter of the blade. You might think you would need to fill that space with blades, but you don't—in large part because the blades spin fast.

Even if the whole area were covered with blades, a turbine can't take all of the energy out of the wind. For that to happen, the air would have to stop completely, and it would accumulate behind the turbine. It turns out, however, that the turbine can extract up to 59% of the energy, provided the turbine is not placed too close to others. That limit is called Betz's law.[8] If we assume the turbine actually reaches the 59% limit, then the power the German turbine can deliver with 20-mph wind is reduced from 10 megawatts to 5.9 megawatts. If the winds get too high, the blades of the propeller are "feathered"; that is, they are rotated about their axes to reduce their "angle of attack" (the amount that they bite the wind). To reduce interference between turbines, they are typically spaced apart by a distance of 5–10 times the blade diameter. Large arrays of turbines

are typically called *wind farms*, although in Texas they prefer the
term *wind ranches*.

Now imagine that you're considering putting a wind turbine on
the roof of your home. You decide to make it pretty big (for a small
roof): about 2 meters × 2 meters, or 6.7 feet × 6.7 feet. That's 4
square meters. Assume that the wind on your house averages 5 mph.
According to our formula, that makes the power density (5 × 5 ×
5)/10 = 12.5 watts per square meter. Multiply that by the area (4
square meters) and by the efficiency (limited to 59%) to get the
output power: 29 watts. That's pretty puny, although it could run
two compact fluorescent bulbs. In contrast, a 4-square-meter solar
cell would deliver 600 watts at peak, and about 75 watts averaged
over cloudy days and 24 hours.

For wind turbines, then, the best prospects are for the very big
ones. New capacity is being installed rapidly. Figure III.8 shows the
installed wind power for five countries. In 2010, the United States
installed about 5 gigawatts peak of new wind capacity (1.7 gigawatts
average, for a 34% capacity factor), bringing the US total wind to
40 gigawatts (peak). In the same year, China installed 15 gigawatts,
bringing its total to 42 gigawatts peak and surpassing the United
States total wind power for the first time. By the end of 2011, the

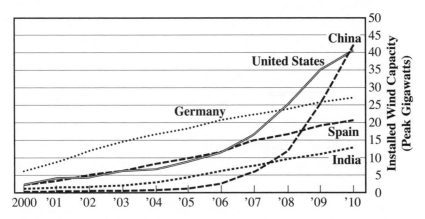

*Figure III.8. Installed peak wind capacity, in gigawatts for five countries.
The most rapid growth is in China and the United States.*

estimated installations in China were 55 gigawatts. The rate of US growth no longer increased, in part because of the recession, but the US capacity still reached 45 gigawatts.

The world wind power capacity has been doubling every 3 years. That rate of growth is astonishing, and it reflects the fact that these huge wind turbines are relatively inexpensive to make, and they don't require fuel. According to Table III.1, wind power installed in the near future could deliver electricity for 9.7¢ per kilowatt-hour. That's comparable to the price of electricity from coal, and cheaper than electricity from nuclear or solar. It is, however, considerably more expensive than natural gas, the killer competitor for all alternative energy.

Wind is only 2.3% of electric power generation in the United States, but it has the potential for rapid growth. Wind turbines require less time to build and install than nuclear power plants, although the rate of growth in the past has been limited by our manufacturing capacity for the gears and generators that make up the hub of the wind turbine. The US Department of Energy would like wind to deliver 20% of our power by 2030. To do that requires continued growth rate of 12% per year. Even in the recession year of 2009–10, US wind capacity still grew by 15%, so such a goal seems within grasp. The problem is that the recent growth has taken place largely in Texas and California, areas in which the high winds are located close to population centers. Growth in the future will have to come from wind in the Great Plains, and that will happen only if the electric power can be delivered to distant cities, requiring an extensive upgrade to the US power grid system. Offshore wind is strong, but construction costs there are high, and the underwater cables to bring the electricity to land are also expensive; that's all reflected in the high costs shown in Table III.1.

Let's discuss some of the common issues that are brought up in discussions of wind power.

What happens when the wind stops?
There are two different answers to this question:

1. As the grid gets larger and there are more wind farms, reliability is somewhat improved. That's because when wind stops at one location, the loss can be compensated for by wind at another. Farms separated by more than a few hundred miles are uncorrelated. There is still a danger from large regional storms if the network is not sufficiently large; for example, in 2009 an extensive storm in Texas caused a major shutdown of wind power.

2. Use backups when the wind stops. When I discuss energy storage in Chapter 10, I'll talk about batteries and compressed air, but the most obvious backup is natural gas. An emergency generator can be turned on in under 10 minutes, and wind is more predictable than that. This raises a secondary question: Natural gas is currently cheaper than wind. So why bother with wind? One answer: to reduce carbon dioxide.

Wind turbines are ugly and noisy

Some people think wind turbines are unattractive; others think they're graceful and elegant. It's a matter of taste, and taste can't be disputed. A relevant comparison is the Eiffel Tower. When it was built as a temporary structure for the 1889 World's Fair, many Parisians found it ugly and afterward were horrified that the structure was not demolished as had been promised.

Here's another quirk about beauty. Some utilities have been running wind turbines *backward* on windless days, supplying electricity to keep the blades moving. They do this for public relations: they've found that many people find a gracefully rotating wind turbine more attractive than a stationary one! (What about you?)

Already environmentalists and nearby property owners have opposed the construction of wind turbines off the coast of Cape Cod. The Alliance to Protect Nantucket Sound says the project violates the Endangered Species Act, the Migratory Bird Treaty Act, and the National Environmental Policy Act. The battle is being fought in the courts.

This is a broad issue that anyone in favor of alternative energy must face. Just the fact that your ideas are supported by some environmentalists, often people most concerned about carbon dioxide and global warming, doesn't mean that if your project goes ahead there won't be other environmentalists who will oppose it for other reasons. Indeed, some people believe the problem is that we already use far more energy than we really need, and that no new power sources of any kind should be built.

Wind turbines kill birds

It's true that turbines kill birds, although currently the number is minuscule compared to the bird deaths caused by other man-made structures, such as tall buildings. I mentioned in Chapter 2 that the US Fish and Wildlife Services estimates that building windows in the United States kill between 100 million and 1 billion birds each year. Moreover, unlike the early wind turbines, modern turbines are normally sited away from bird migration flyways.

Isn't the wind too remote?

The main problem with wind power is that the strongest wind regions are far from the population. Delivering the power will require a modernized electric power grid. The current US grid wastes about 7% of the energy in the electricity it transports; long wires would be worse. But that loss is due to heating in the wires; it can easily be avoided by using fatter transmission lines. The old ones were built when electricity was cheap and extra aluminum for thicker wires was not. Yet the major difficulty in improving the electric grid is not in replacing the wiring, but in obtaining the right of way to build the new towers and string the wires over the land in between. This is such a sticky legal issue that many people assume new construction can only be built along existing routes. Some people have argued that any new economic stimulus money would be better spent on an expanded power grid (and perhaps in making it smart) to enhance the possible future growth of wind power, rather than on expensive technologies such as solar that may always require subsidies.

10

ENERGY STORAGE

CAN YOU save solar energy for a rainy day? Or just for a dark minute, if a cloud passes over your home solar cells? The answer is yes; I already mentioned that the heat from solar-thermal plants can be saved. But what about the electricity from solar cells and wind turbines? Again, the answer is yes; there are almost as many ways to store energy as there are to generate it, including batteries, compressed air, flywheels, ultracapacitors, and even hydrogen fuel generated from water by electrolysis. We'll discuss each of these, but let's start with the probable winner: batteries.

Batteries

Batteries are little chemical laboratories that use fuel to separate electrons from their atoms. Let the electrons return through a wire, and you have electricity. I'll give more details in a moment, but first let's appreciate what miracles batteries are. Pump energy into a lead-acid automobile battery by charging it up, and when you draw the power out you'll get 80%–90% of it back. That's an amazing efficiency, and it's the reason that many solar cells for the home have lead-acid backup batteries included in the price. Four car batteries weighing 250 pounds in total can store 5 kilowatt-hours of electricity, enough to run a small home for 5 hours. That efficiency is amazing, but the downside is that the energy stored is very small compared to the energy available in ordinary fuels. With gasoline, 250 pounds provides 1,320 kilowatt-hours of heat energy—263 times more than car batteries. With a generator that turns that into electricity at only 20% efficiency, gasoline still provides about 50 times the electric energy of an equal-weight battery.

Even though their energy density is low, batteries are an efficient way to store energy, particularly if you have plenty of space. For solar and wind farms, the common lead-acid battery is not the obvious choice; a stronger contender is the sodium-sulfur battery. The largest sodium-sulfur battery yet installed is located in Presidio, Texas, and is named *Bob*. (Some people claim "Bob" is an acronym for "big old battery.") The photo of Bob's home in Figure III.9 gives a sense of its size.

Actually, Bob is not used as backup for solar and wind, but as an emergency measure in case (or rather, when) the single power line connecting Presidio to the US power grid fails. Bob can provide 4 megawatts, enough to supply power for 4,000 homes, for 8 hours. But sodium-sulfur batteries are also being used for solar and wind backup, and for power "leveling"—keeping the line power constant in the face of generator variations. Duke Energy plans to install a 36-megawatt sodium-sulfur battery (made by a start-up named Xtreme Power) at its 153-megawatt Notrees Windpower Project in Texas.

The advantage of sodium-sulfur over other batteries is the price per charge-discharge cycle. Current sodium-sulfur batteries can be recharged 4,500 times (for 80% discharge), versus only 500 for typical lead-acid and lithium-ion batteries (although laboratory models do better). That's a 9-fold advantage. I expect that lithium-ion batteries will *never* be used for large-scale energy storage; they are too expensive. Lithium costs 40 times more per pound than sodium, and 10 times more per atom—a more relevant measure for batteries.

Figure III.9. The building in Texas that holds Bob, a very big battery for energy storage.

With the 9-fold recharge advantage and the 10-fold cost-per-atom advantage, sodium-sulfur has a 90-fold advantage over lithium-ion.

Why, then, don't we use sodium-sulfur batteries in our cell phones and tablet computers? The catch is that these batteries require liquid sodium and thus work only at high temperatures, typically about 660°F (350°C), hotter than your home oven. Such temperatures are not a serious problem for commercial applications. Sumitomo Electric Industries has announced that it hopes to reduce the operating temperature to lower than 212°F (100°C), the boiling point of water, to make the batteries usable in buildings and maybe even large vehicles, such as buses—but even then, not in laptops

THE PHYSICS AND CHEMISTRY OF BATTERIES

Future presidents can skip this section on how batteries work, but I find it very interesting and maybe you will too. Batteries take advantage of the unusual properties of two kinds of materials: metals and electrolytes. Metals freely conduct electrons; that's why we make wires out of metals. More mysterious are the electrolytes, materials that don't conduct electrons but do allow atoms to flow through them. In a lead-acid auto battery, lead and its compounds act as the metal,[9] and the acid-water mixture is the electrolyte. The key trick is to have positively charged atoms (called *positive ions*) drift through the electrolyte and then become chemically stuck on the other side. They attract the electrons they left behind, but since the electrons can't move through the electrolyte, they have to take a roundabout way through a wire that you provide. While the electrons are flowing through this wire, you can extract their energy.

There are many electrolytes to choose from (salt water works, as well as the flesh of potatoes) and lots of materials for the metals. These have been surveyed and analyzed, and virtually every combination you can imagine is described in *The Handbook of Battery Materials*.[10] The real trick with batteries is figuring out how to make them rechargeable. To recharge a battery, you use a generator to force the electrons to return to their original side; when there, their negative charge will attract the positive ions to break away from the compounds that they stuck to and drift back through the electrolyte. That's a great idea, but the difficulties are in the details. The ions must go back to the electrode and attach themselves in a benign way, ideally back to the same configuration that they had before they left. They often don't; a persistent problem with rechargeable batteries is that the returning ions tend to form long fingerlike structures called dendrites. If the dendrites grow with each recharge cycle, they may eventually make the battery unusable. Typical rechargeable batteries fail after a few hundred recharges. Part of the interest in sodium-sulfur batteries is their ability to be recharged thousands of times without failure.

THE FUTURE OF BATTERIES

Over the past few years, we've seen a dramatic improvement in the batteries that we use. Not long ago we used nickel-cadmium batteries—NiCad's—and we hated them because of the troublesome memory effect. (They had to be drained completely before being recharged.) Then we had nickel–metal hydride batteries—NiMH's; these are still used in the Toyota Prius. We felt blessed—the memory effect was gone! Then came lithium-ion batteries, with their light weight and high energy density. Next lithium polymer batteries, made so thin that they could be hidden in your cell phone, your e-reader, or your now impossibly thin laptop. These rapid developments have led many to an optimism that batteries, just like computers, will continue to get better at a dizzying pace.

In my estimation, such optimism is unwarranted. What seemed like a rapid development of the battery technology was, in reality, just a rapid development in the battery *market*. Just two decades ago, nobody was willing to pay $100 for a pound of battery. That all changed with the laptop/cell phone/digital camera revolution. If you have a $1,000 laptop, then you might be willing to buy a $100 battery to keep it operating. So the presence of this new market meant that the wide range of battery technologies already known, those in the handbooks, could be commercially developed. Most of the work of the past few years has been to address the messy technical details of rechargeability and safety. This is the kind of engineering development that is linear, not exponential. So expect batteries to improve, but not at the pace that we've seen in the recent past.

Bottled Wind: Compressed-Air Energy Storage (CAES)

Air can be readily compressed to several hundred times atmospheric pressure. That gives it an energy storage capacity per volume comparable to that of batteries. The energy is readily retrieved by running the air through a turbine (that is, a fancy fan). I've ridden on a compressed-air vehicle in a gold mine; in that confined space, with poor ventilation, using gasoline was inadvisable, and compressed air was considered safer than lead-acid batteries—which, of course, contain sulfuric acid. Moreover, unlike batteries (which last only 500 to a few thousand cycles), compressed air tanks can be used and reused virtually indefinitely. You force the air in using a pump, often a piston or a turbine. This takes energy, typically from an electric motor, and that energy is the energy you're storing.

One problem is the weight of the tank. It is a remarkable engineering fact that, regardless of size, a steel tank holding compressed air will weigh about 20 times more than the air itself; a modern fiber-composite tank will weigh 5 times as much as the air. So you don't get a weight advantage by using one large tank versus several smaller ones. The reason for this surprising result is that a larger tank requires thicker walls to hold itself together against the force of the compressed air.[11]

Another problem with compressed-air energy storage (CAES) is that the air heats up when you compress it. That can be a big factor; if the heat can't flow away, then when you pressurize to 200 atmospheres (a typical value), the gas temperature will rise to nearly 1,370°C (2,500°F)![12] On the other hand, if you let that heat escape (for example, you let the tank return to room temperature), you lose a good fraction of the energy you put in. You can get that back if you release the pressurized gas slowly enough, since as the gas expands it cools, and it will absorb heat from the surrounding environment.[13]

Currently, not many CAES systems are in operation. There is one in Huntorf, Germany, and another in McIntosh, Alabama; neither uses a metal tank. The Alabama facility puts compressed air in a cavern that was hollowed out of a salt dome (a solid underground salt deposit) by flushing water through it. The cavity is 900 feet long and 238 feet wide. A new plant planned for an abandoned limestone mine near Norton, Ohio, will be able to deliver 2.7 gigawatts. Other CAES projects have been designed for California, New Jersey, and New York. In the advanced designs, the heat generated by compression is removed and stored, and then used to heat the gas again when it is expanded to run in a turbine. Calculations show that with such an advanced system, we may be able to recover as much as 80% of the energy pumped underground, comparable to the recovery you get from a battery. This kind of "adiabatic" CAES project is also planned to begin in Stassfurt, Germany, in 2013.

Although manufactured tanks can be used for small vehicles (such as the gold mine cart I rode), for city-scale CAES the cost of the tanks is too high. The cavities have to be geologic. According to the Department of Energy, suitable sites exist over much of the Midwest, where the wind is also abundant.

Many people are optimistic about compressed-air energy storage. The Electric Power Research Institute (EPRI) predicts that CAES will be an important part of our energy future. The ultimate fate of CAES may be determined by the competition of natural gas, and by whether there is a financial incentive to reduce carbon emissions.

Flywheels

Spin a wheel using a motor, and you've stored kinetic energy in the rotational motion. Make the wheel heavy and spin it fast, and it can store a lot of energy; such a wheel is called a *flywheel*. If you're clever, you use a motor that can act in reverse—that is, become a generator. Spin it up, and then detach motor power leads from the energy source and attach them to a lightbulb. The energy of the spinning flywheel will now cause the motor to generate electricity; the bulb will light and the rotation will slow as the kinetic energy is converted to electrical.

A flywheel is not only a way to store energy, but it is a useful way to *condition* energy delivery, to even out the fluctuations. A flywheel can deliver its energy very rapidly when called upon. At the Lawrence Berkeley National Laboratory, where I did most of my physics research over the years, the atom smasher called the Bevatron had multi-ton flywheels to smooth out the power it took from the grid. The Bevatron needed energy for only a brief time every 6 seconds, and without the flywheel its draw would have made the lights of Berkeley dim every 6 seconds. The flywheel took in energy when the Bevatron didn't need it, and delivered it as a supplement when the Bevatron hit its maximum need.

A fascinating anecdote about the Bevatron flywheels was told to me by Nobel Laureate Luis Alvarez, who had participated in their design. These huge flywheels, about 10 tons each, were carefully oriented so that if they ever broke loose (an event that was not expected, but they *were* situated near the Hayward Fault), then after they smashed their way through the walls of the Bevatron building they would roll away from the city of Berkeley and up over the hills toward the reservoir. This scenario illustrates the potential problem with all energy storage systems (not just flywheels): safety.

Modern flywheels don't look at all like those giant wheels at the Bevatron; they look more like tubes, similar to the centrifuges used for uranium enrichment. An example is shown in Figure III.10. Why are they tall and narrow? The reason comes down to some basic physics and materials science. Energy is stored in the velocity of

Figure III.10. Beacon Power "Smart Energy" 25-inch-diameter flywheel.

motion of the flywheel material (typically high-strength steel or a carbon-fiber composite). For optimum use, such material should be in the form of a hoop, so that most of the material is moving at the top speed. (A wheel has slower velocity close to the hub.) The speed at which the hoop rotates is limited by the strength of the hoop. A calculation shows that the maximum velocity of the hoop is independent of the hoop radius. (This is a good exercise for an undergraduate physics major.)[14] The result is that you can use space more effectively by using small hoops and placing them close to each other. Of course, the hoops can also be stacked, making cylinders.

The 2,500-pound carbon-fiber composites in Beacon Power "Smart Energy" flywheels twirl around at 1,500 miles per hour. That's Mach 2! To reduce supersonic friction with air, the chamber holding the flywheel is pumped to a high vacuum. Each cylinder can store 25 kilowatt-hours of energy. Beacon Power recently installed an array of 200 of these flywheels in Stephentown, New York, capable of storing 5 megawatt-hours of energy. The flywheels are designed to deliver 20 megawatts, and that means they can run for ¼ hour = 15 minutes. That doesn't seem like a lot, and it isn't. These flywheels are being used for the traditional purpose: power regulation. They help keep the power of the local grid at constant frequency, despite a rapidly changing load.

Will flywheels ever be used for large-scale energy storage—for example, for a wind or solar farm? The kinetic energy at 1,500 mph is about 30 watt-hours per pound—comparable to that of a lithium-ion battery. That makes them seem attractive. But only a third of the weight of the system is in the spinning flywheel; the rest is in the vacuum vessel. And most of the space in the flywheel structure is empty. The current flywheels (including the vacuum vessel) are about 10 feet tall and 6 feet in diameter. That means they store only 2.6 watt-hours per liter. In contrast, a lead-acid battery holds 40 watt-hours per liter. Beacon Power's current system can deliver energy for about $1.30 per kilowatt-hour—very expensive compared to the average US wall-plug price of 10¢. The design is very sophisticated, and the price is unlikely to come down very much. As a result, I would guess that flywheels will continue to be used for power conditioning, but not for large-scale energy storage for wind or solar farms.

Supercapacitors

A capacitor is a set of two metal surfaces separated by an electric insulator. Put positive electric charge on one plate, negative on the other, and the combination can store energy for a long time, much longer than batteries can. Add more electric charge on the plate and you store more energy, but you also raise the voltage. Keep adding charge, and eventually the increasing voltage will cause electric breakdown, a spark that could permanently damage the capacitor. The trick for storing energy in capacitors is making the insulator very thin, so that you can have lots of energy per unit volume while keeping the voltage low. Unlike batteries, capacitors don't depend on chemical reactions, so they can release their energy extremely quickly, and they don't degrade with use and time, at least not as rapidly as rechargeable batteries do.

Just in the last few decades there has been an astonishing development in capacitors. The new high-energy-density capacitors are called, naturally, *supercapacitors*, or sometimes *ultracapacitors*. They are also sometimes called EDLCs, for "electric double-layer capacitors," but that is a much more boring name. Supercapacitors can store as much as 14 watt-hours per pound, about a third the energy of a same-weight lithium-ion batteries, but they currently cost over 3 times as much. That's a nine-fold disadvantage in energy stored per dollar.

Their main value will probably be for use in combination with an ordinary battery. Battery lifetime is significantly hurt when batteries have to provide short, intense bursts of energy, but this is precisely what a supercapacitor can do easily. Because they can be charged so quickly, supercapacitors can be used to improve the efficiency of regenerative braking; they absorb the energy and then can transfer it at a more leisurely and efficient pace to the battery. But in my estimation, supercapacitors will not be major contributors to our large energy storage needs.

Hydrogen and Fuel Cells

Fuel cells have a romance about them, perhaps born in the space program when the public first became aware of them. You'll see a lot of fantastic claims about their future potential—a lot of optimism bias. They do have true value, and important applications, but they will not be a general replacement for either batteries or generators.

A *fuel cell* is basically a battery that doesn't have to be recharged. Instead, you simply replace the chemicals that provide the energy. For a hydrogen fuel cell, you pump in hydrogen and air, and out comes electricity. But for energy storage, you have to also operate them in reverse, to produce the fuel. Unfortunately, this process has low efficiency, typically only 25%. Compare that to the 80%–90% that batteries provide. Fuel cells may find a role as a substitute for a turbine in primary energy generation. Some people think they can replace the motor in an automobile. I'll defer the discussion of these issues to Chapter 16.

Natural Gas

Natural gas is not an energy storage method, but it is still the main competitor to energy storage, so it is appropriate to consider some relevant numbers here. Let's compare a natural-gas generator to a top energy storage technology: sodium-sulfur batteries.

First, battery cost: A sodium-sulfur battery that can store 1 kilowatt-hour of electricity costs about $500; that's 50¢ per watt-hour. If you want it to run for 10 hours (during a wind lull or on a dark day), then it can operate at only 100 watts. The capital cost can be expressed as $5 per deliverable watt.

Next, natural gas cost: In the discussion of blackouts (see Chapter 7), I gave the example of a peak natural-gas generator costing $100 million for 100 megawatts; that's $1 per deliverable watt. Sodium-sulfur batteries are 5 times more expensive. Natural gas wins easily. If we include the cost of the fuel, then natural gas wins even bigger, since it is the cheapest source of energy, much cheaper than the solar or wind power that presumably is used to charge the batteries.

Why ever use batteries? Was Bob (the big old battery in Texas) a mistake? Batteries certainly are simpler in operation and maintenance; they can start up instantly (natural gas takes a few minutes), and in some locations natural gas is not conveniently available. For a small facility, the ease of battery use may be decisive. If you run with a very low-duty cycle—say, for just 1 hour per day rather than 10—then the per-watt capital cost of the battery drops from $5 to 50¢, and batteries become economically competitive. But you always need to look carefully. Whenever you're thinking seriously about alternative energy—any kind—you need to remember that from an economic perspective, natural gas is the competitor to beat.

11

THE COMING EXPLOSION OF
NUCLEAR POWER

I BEGIN with an executive summary, a list of some key things you need to know about nuclear power. The items on the list were chosen because each is important, and because each will surprise most non-experts. Several of these were discussed previously in this book, but here we'll go into more depth.

- **Blowing up.** Nuclear power plants *cannot* blow up like atomic bombs— not under any circumstances, even if a terrorist with a PhD in nuclear physics were to gain complete control. That's because these power plants use "low enriched uranium," which has very different properties from the "highly enriched uranium" required by nuclear weapons.
- **Cost.** Nuclear reactors are expensive, but that doesn't mean the cost of electricity they produce is high. The big expense is in building them— the capital cost—but their fuel and maintenance costs are very low. Once the loan to build them has been paid off, they provide some of the cheapest power available.
- **Medium size.** New "medium-sized" nuclear plants significantly reduce the initial investment and the need for government loan guarantees. They also improve safety by being naturally immune to the kinds of accidents that have caused havoc in the past.

- **Running out of uranium?** We are not about to run out of uranium for fuel. There is enough economically recoverable uranium to last (at current usage rates) for 9,000 years. We *are* running out of *cheap* uranium, but the cost of the uranium ore for 1 kilowatt-hour of electricity is about 0.2¢. Even if its price skyrocketed, uranium would remain only a small component of the electricity cost.
- **Fukushima deaths.** Of the 15,000 deaths from the 2011 tsunami, only 100 of them will come from the Fukushima nuclear accident—and maybe fewer, since thyroid cancer is readily treatable.
- **Nuclear waste storage.** Storing nuclear waste is not a difficult technical problem. It has been solved. It is more a problem of public perception and political posturing.
- **The coming nuclear explosion.** Regardless of whether the United States develops new nuclear plants, the rest of the world is forging ahead. There are major programs in China and France, and even Japan, currently turning off its own nuclear power, hopes to manufacture nuclear reactors for the rest of the world. This *explosion of use* is what gave this chapter its title.

Let's look at each item in more detail.

Blowing Up

Can a nuclear reactor explode like an atomic bomb? I discussed this briefly in Chapter 1, but here I'll go through it in detail. Both atomic bombs and nuclear power reactors are based on the chain reaction of uranium or plutonium. When a nucleus of uranium splits, it releases as heat 20 million times more energy than does a molecule of TNT, and in the process spits out two or three fast particles called neutrons. When these neutrons hit other uranium atoms, they make those atoms fission too. Just for simple math, assume that the number of neutrons released per fission is 2. The first 2 neutrons trigger 2 additional fissions, which together emit 4 neutrons that trigger 4 more fissions, and so on. What starts out as one fission soon becomes

2—and then 4, 8, 16, 32, 64, 128, 256, 512, 1,024, and so on. This doubling process is what we call a *chain reaction*. You can work out the math in a spreadsheet (if you use spreadsheets in your work, try it!) but here's a shortcut: every 10 doublings increases the number by a factor of 1,024, about a thousand. So after 80 doublings, the number of neutrons is roughly 1,000 multiplied by itself 8 times. That's 1,000 × 1,000 × . . . × 1,000 with 8 factors, equal to 1 followed by 24 zeros.

Let's apply this math to a bomb design. Suppose we want a bomb as powerful as the one that destroyed Hiroshima—13,000 tons of TNT equivalent. How much uranium do we need? Since uranium releases 20 million times more energy than TNT, the amount we need is 20 million times smaller: 0.00065 tons of uranium, equal to 1.4 pounds (0.65 kilogram). The Hiroshima bomb needed much more than this in order to have a *critical mass*, one that would actually explode, but 1.4 pounds is all that actually fissioned. From these numbers, we can show that the number of doublings it takes to explode every atom is indeed about 80.[15]

It's that small number of doublings—80—that makes the bomb possible. Suppose each doubling takes 10 billionths of a second. (Ten billionths of a second is about the time it takes your laptop computer to do 10 computations. In the nuclear business it is called a *shake*, originally named after the rapidity of a lamb's tail.[16]) Then all the atoms will be split in just 800 billionths of a second, less than a millionth of a second. That short time is essential in nuclear bomb design. If the number had turned out to be a full second (instead of one millionth) then the initial fissions would have released enough energy to blow away most of the uranium atoms before they had a chance to fission. That's called *predetonation*. The quickness of the chain reaction is absolutely essential to avoid predetonation.

I didn't mention an important fact: ordinary (heavy) uranium, called U-238, does not fission in a way that can sustain a chain reaction. Only the light U-235 can be used.[17] Here's the problem. Because U-238 is so abundant, most of the neutrons will hit it first and be absorbed, turning the U-238 into U-239 and eventually into plutonium. That means the chain reaction does not work for ordinary uranium. The neutrons are eaten up.

In contrast, when light U-235 is hit by a neutron, it almost always fissions, releasing more neutrons. To make a bomb, you need to get rid of the heavy U-238 pollutant. The uranium must be enriched to nearly 100% U-235 for a bomb to work. You can't have very much of the pollutant U-238 around, or it will absorb too many neutrons and fizzle the chain reaction. Unfortunately for the bomb designers (but maybe good for humanity), U-235 accounts for only 0.7% of ordinary uranium, and it is very difficult to separate. Purification, called uranium *enrichment*, was accomplished by a group of scientists during World War II in the Manhattan Project. Uranium enrichment is the purpose of the centrifuges being built in Iran and North Korea. The world worries that the Iranians and the North Koreans plan to enrich U-235 from its natural 0.7% level, to *weapons grade*—90% or higher.

If you want a chain reaction but you're not building a bomb, there's a trick that lets you use impure uranium. It takes advantage of the physics discovery that *slow* neutrons are not readily absorbed by U-238; they tend just to bounce off. It's the fast ones that stick. That discovery was surprising, but very useful for nuclear reactor designers. If you do something that slows the neutrons, you can get a good chain reaction even if a lot of U-238 is present. Two materials that can do this slowing are water and carbon. Put in enough of these (typically much more than the amount of uranium), and the neutrons bounce off the slowing molecules and lose energy and speed. If they slow down enough before they hit a U-238 atom, they will not be absorbed, and they will continue bouncing around until they find U-235. Materials that slow neutrons are called *moderators*.

If you use an expensive moderator (heavy water), then your nuclear reactor can work with natural uranium, which contains only 0.7% U-235. A graphite moderator can also work with natural uranium, but graphite poses a serious fire danger; the burning graphite at Chernobyl exacerbated that disaster. If you use ordinary water—a cheap and relatively safe moderator—then you need to enrich the uranium, but only to 3% or 4% U-235. The Canadian CANDU reactor uses heavy water; the Chernobyl reactor used graphite as its primary moderator;[18] the Fukushima reactor and the power reactors in the United States use ordinary water.[19]

If the moderator is lost (for example, the water leaks out), the neutrons won't slow, so they'll be absorbed by U-238; the chain reaction stops. The neutrons must be slowed to keep the reaction going, but slowing the neutrons has an important safety implication. Remember that about 80 doubling stages are needed for the full power of the atomic bomb. But after only 60 stages, the energy release already equals that of TNT. So after 60 stages, the whole core of the reactor should blow up—with TNT-like energy—and when that happens uranium is dispersed and the chain reaction stops.

If the uranium blows apart as soon as it reaches TNT energy density, how can a nuclear bomb release 20 million times more? The trick is that for a bomb you don't use a moderator. Since the neutrons are fast, the stages go much quicker—in fact, about 7,000 times quicker. The released energy grows and reaches that of TNT and then surpasses it, and the bomb starts to explode—but the neutrons are going faster than the explosion. As the uranium begins to fly apart, the *fast* chain reaction from the *fast* neutrons continues, completing all 80 stages. The neutrons outrun the explosion. Of course, if you're going to operate without a moderator, you need to get rid of most of the U-238. You need weapons-grade uranium with less than 10% contamination from U-238.

The Chernobyl nuclear reactor used impure uranium, so it incorporated a moderator to slow the neutrons. As a result, when the chain reaction ran out of control, the reactor blew up, but only at a TNT level of explosion. The explosion destroyed the reactor building and did some nearby damage, but it was the release of the nuclear waste, not the explosion, that caused virtually all of the deaths.

Cost

Nuclear power has a reputation of being the most expensive kind of traditional electric power production. This perception is based in large part on the simple observation of the high capital cost it takes to build and install 1 gigawatt of capacity: currently between $6 billion and $8 billion. That's about 50% more than it costs to build a 1-gigawatt coal plant, and over 4 times more than the cost to build a conventional natural-gas plant. The high cost is a result of the complexity of nuclear power plants and the high quality control needed in their construction. With such a high price it seems absurd to go nuclear—so why are nuclear power plants still being built? Is there a conspiracy? Maybe there are subsidies; maybe the nuclear lobby donates lots of money to politicians; maybe . . .

The calculation changes, however, when you include fuel and operating costs. After the plant is built, the incremental cost of electricity from nuclear power is cheaper than from any other source except hydroelectric dams. According to the US Energy Information Administration, 80% of the cost of electricity from nuclear power comes from paying back the loans (with interest) for the construction cost, the capital investment. In contrast, for a natural-gas power plant only 18% of the cost is for capital investment; the rest is the cost of the fuel. For natural gas at US prices, the net cost is still lower than for nuclear, but in other places around the world, nuclear is competitive.

Nuclear wasn't always this cheap. Over the past 30 years the plant *capacity factor*—the fraction of the time the plant is operating and delivering power—rose from 55% to nearly 90%.[20] The improved capacity factor makes a big difference to an investor. If the plant operates only half of the time, then you're earning money only half of the time. Today, with plants operating 90% of the time instead of 55%, yearly revenues are 1.6 times larger. This improvement is primarily due to better operating procedures, use of modern computers to monitor and control the plant, and a great deal of experience— regularly shared among reactor operators at meetings and workshops.

Back in the 1970s, nuclear power surged; in 1973 alone, US utilities ordered 41 new plants. Then, in 1979, the Three Mile Island accident occurred. Although the Kemeny Commission that studied the accident determined that nobody had been killed, it did report some health damage to the public—by psychological stress on an unnecessarily frightened public who thought the accident was much worse than it really was, perhaps because they had recently seen the movie *The China Syndrome* in theaters, a movie that assumed incorrectly that reactors could blow up like atomic bombs. Orders for new nuclear reactors came to a halt. Then, in 1986, a truly catastrophic reactor accident occurred at Chernobyl, causing 30 prompt deaths (mostly firefighters),[21] an estimated 24,000 long-term cancers from the released radioactivity, and an evacuation of the surrounding region.

Nuclear power continued to provide about 20% of US electric power, but no new plants were ordered. A few, already in the pipeline, were built and commissioned. Nuclear power was pronounced dead by its opponents, including the Sierra Club and Ralph Nader.

This death of nuclear power growth wasn't due only to public fear. Between 1984 and 1986, the price of natural gas plummeted by 50%. Economically, nuclear couldn't compete, at least in the United States. But like a cat, nuclear power seems to have many lives. Some countries (France, Japan) went heavily nuclear because they had little or no known fossil fuel resources. The improvement in capacity factor at nuclear power plants meant that the yearly energy output continued to grow, even while the number of plants didn't.

Remarkably, history has repeated itself. The cost of natural gas had been about $7 per thousand cubic feet from 2004 to 2008. (It shot up twice, in 2005 and 2008, to over $13.) But in 2009 it plummeted to $4 per thousand cubic feet, thanks largely to the development of economical methods to extract it from the huge shale gas reserves, and I expect it to remain this cheap or cheaper for the foreseeable future. In Europe and in Japan, where natural-gas prices have so far remained above $11 per thousand cubic feet, nuclear is still very competitive.

Small Modular Nuclear Reactors

The cost of a nuclear plant is so high that few companies can afford to take the risk of building one. The greatest uncertainty may be that public outrage, perhaps over some new event (Three Mile Island, Chernobyl, Fukushima), will delay construction. Delays can be very expensive as loans continue to accumulate interest. Because of this risk, many companies can't obtain financing for new plants without a guarantee from the US government. Many people consider such loan guarantees a "subsidy" of nuclear power, although in fact the government has never actually had to pay any money for bad nuclear loans (unlike the case with the solar-cell manufacturer Solyndra, discussed in Chapter 8).

A new way around this financial barrier is to build *small modular reactors*, power plants that produce only 300 megawatts of electric power or less, and which require a smaller capital investment. In the past, such small reactors were thought to be impractical. Even little ones require elaborate and expensive safety systems. But advanced (generation III and IV) reactor design solved these problems. These reactors are called modular because several can be used together at the same location, sitting next to each other. If your power requirements grow, you can simply add more modules.

Several companies have now designed what appear to be practical small modular reactors. One advanced design by Babcock & Wilcox (who manufactured the reactor at the Three Mile Island plant), produces 125 megawatts, one-eighth of the typical prior 1-gigawatt size. The entire reactor is constructed in a factory, shipped by railroad to the reactor site, and then buried underground. It can operate for 3–4 years without maintenance.

The 4S reactor designed by Toshiba (4S stands for "Super-Safe, Small and Simple"), pictured in Figure III.11, will deliver 30–135 megawatts. The 4S uses liquid sodium coolant to take away the heat energy used to power a turbine and generate electricity. The reactor is designed to be buried. It is sometimes referred to as a "nuclear battery" because of the way it operates, requiring little attention, even when you draw no power. The only control for the reactor is a series of reflectors around the sides of the reactor cylinder that can be moved in or out. If these reflectors are removed, the reactor chain reaction stops; if left in place, they control the level of the chain reaction through physics, with no physical motion required.

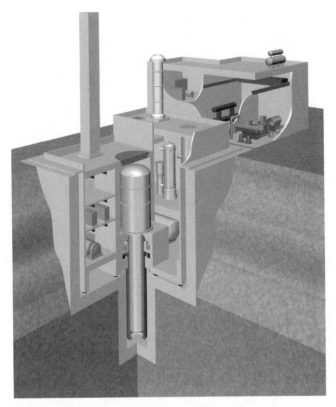

Figure III.11. The 4S small modular nuclear reactor.

Because the coolant in the 4S is liquid sodium, a metal, it can be pumped through the reactor using electromagnetic pumps—which have no moving parts. If the pumps fail, the reactor will begin to overheat, but the properties of the materials in the reactor will reduce the chain reaction. No operator intervention is necessary; no components will even have to physically move. Excess heat will be removed by natural convection in the sodium. Thus these reactors are designed to be intrinsically safe, depending only on the physics of neutron reflection and convection. No engineering systems are required to function for the purpose of safety.

A potentially worrisome issue is that unlike the large reactors, small modular reactors do not moderate the neutrons; that is, the neutrons in the chain reaction are fast. Allowing the neutrons to maintain their high speed is the fundamental trick employed to make these reactors smaller. But to use fast neutrons, the reactors can't have too much U-238, the contaminant that absorbs fast neutrons. So the operators of these reactors must use uranium fuel enriched to 19.9% U-235, not quite bomb quality, but much higher than the 3%–4% used in the larger reactors. Keeping the number below 20% allows them to call it "low enrichment" by the standards of the International Atomic Energy Agency.

Recall that it's the use of slow neutrons that prevents a large reactor from exploding like an atomic bomb. The small modular reactors use a different trick. To keep the chain reaction going, on average one neutron from each fission must hit another U-235 atom.[22] As atoms heat up, they vibrate more rapidly (that's what heat is), and the increased vibration makes them move farther apart (that's why things expand when they get hot). The increased space between the hot atoms gives the neutrons an increased chance of escape. The reactor core has become a leaky bucket. Once its probability of one of the neutrons hitting another U-235 atom drops below 1, the chain reaction quickly shuts itself off. If it is, for example, 0.999, then after 100,000 stages (which takes only a thousandth of a second), the number of neutrons has dropped to 0.999 multiplied by itself 100,000 times. That's 4×10^{-44}, or 0.000000000000000000000000000000000000000000004. This number is so tiny[23] that it means that there are no more neutrons being produced; the chain reaction has stopped.

Note that the safety is not based on an engineering system that requires maintenance; it is not one that can be subverted by a terrorist or a confused technician who turns off the wrong valve. The safety is intrinsic to the physics of high temperatures. That's why these reactors are sometimes called *intrinsically safe*.

There's a second physics reason why these reactors are intrinsically safe. When the fuel heats up, both the neutrons and the uranium atoms shake more; their instantaneous velocity is higher. U-238 has an important property in its nucleus: it becomes more efficient at absorbing neutrons when the relative velocity is increased. So, hot U-238 is a better neutron absorber than is warm U-238. If it steals enough of the neutrons from U-235—and just a little bit of theft is enough—the chain reaction stops.[24]

Of course, safety is not just a matter of preventing a runaway nuclear chain reaction (such as in the reactivity accident that occurred at Chernobyl), but also a matter of protecting the reactor against a loss-of-coolant accident, such as occurred at Fukushima and at Three Mile Island. In those accidents, even though the chain reaction was shut off, the continued heat from radioactive decay, although it dropped rapidly, was sufficient to melt the fuel when the external pumps failed to bring cooling water into the reactor core.

In these small modular reactors, the cooling is designed to work even without a pump or external power. The cooling is based on convection. The hot fluid in contact with the fuel (for example, liquid sodium) expands when heated; we already discussed how the expansion slows the chain reaction. But the expansion also makes the liquid less dense—that is, lighter than an equal amount of the surrounding fluid. Because it is lighter, the liquid rises like hot air, carrying its heat away from the core; cooler liquid replaces it. This natural flow does not depend on pumps or engineering devices, but only on the physics of materials. So in this aspect, too, modular reactors are intrinsically safe.

Some people argue that another advantage to these reactors is that the waste contains less plutonium than in conventional reactors. Plutonium is produced in all reactors by the neutrons that stick to the U-238 (even if the neutrons are slow, not all of them will bounce off). But the modular reactors are designed to burn fuel for very long times without change of fuel, on the order of 5–30 years. That means that much of the plutonium produced is also fissioned, so there is less of a plutonium excess when the reactor is finally dug up and the spent fuel removed. But I don't consider such plutonium to be a major problem in the current reactors; it is more of a political problem based on the public's misguided and exaggerated fear of plutonium. I think it is best to address misplaced concerns through education.

Because of their size and intrinsic safety, small modular reactors also offer the opportunity to use nuclear power in locations that are not part of a large electric grid system—for example, in many cities in the developing world. Some people worry that putting nuclear reactors in remote areas or in countries that may not have high security makes them vulnerable to terrorism. Countering that concern is the fact that these reactors are buried underground. The only access to them, needed every 5–30 years for refueling, is to dig them up. That takes considerable effort and expense and time. Burial provides a natural resistance to terrorism.

Another worry is the danger that 19.9% enriched uranium poses to proliferation issues. Once uranium is enriched from its natural 0.7% to this level, the effort to take it to weapons grade (90% U-235) is small. Think of it this way. To obtain 1 kilogram of pure U-235, you have to start with 140 kilograms of natural uranium. If you start with 20% enriched, you need only 5 kilograms. So for the initial stages, you have to have centrifuges that handle much more material—28 times as much! And there's a second factor.[25] Suppose every pass through the centrifuge enriches uranium by a factor of 1.3. Then the number of stages to get from 0.7% to 20%—a factor of 28—takes 100 stages (that's because $1.3^{100} \approx 28$). To go from 20% to 90% is enrichment by a ratio of 4.5, and that takes only 17 stages (because $1.3^{17} \approx 4.5$). So not only are you handling much less material (by a factor of 28) but you require fewer stages of enrichment (by a factor of 100/17 = 5.9). Combined, the extra effort is a factor of the extra uranium you need (a factor of 28) multiplied by the number of stages you have to use (5.9), giving a total ratio of difficulty of 28 × 5.9 = 165.

In simple language, it is 165 times harder to enrich uranium from natural 0.7% to 19.9% than it is to take it the rest of the way, from 19.9% to 90%. This is a serious issue. Once you have 19.9% enriched uranium, you can enrich to bomb-grade by using some primitive technology, such as the calutron. (The calutron was a device for uranium enrichment invented by Ernest Lawrence during World War II and used for the Hiroshima bomb; it was also the initial device used by Saddam Hussein prior to the first Gulf War to obtain partial enrichment.[26])

This danger of a market for 19.9% uranium is certainly a concern, but I feel it may not be a truly serious problem. These reactors are built in the factory, with the uranium an integral part installed during manufacture. Digging up one of these reactors to extract the uranium could not be a clandestine operation; it would be blatantly obvious to the rest of the world. And the material extracted would still have to be enriched to achieve weapons-grade uranium. It is also worth recognizing that even with U-235, you need either an implosion design (very difficult, requiring lots of testing) or a gun design like that used at Hiroshima, in which a cannon inside the bomb blows together two subcritical masses. Such a device is heavy because it includes not just the uranium but also the cannon; the U-235 bomb that destroyed Hiroshima weighed 5 tons. These weapons cannot be delivered in either suitcases or simple missiles; even a large Scud missile can carry only 1 ton. Delivery of such a bomb to a target requires a truck or airplane.

Of course, the United States with its sophistication and experience can construct lighter nuclear bombs—our artillery warhead, the W33, weighs only 243 pounds. But such devices are exceedingly difficult to design and fabricate, particularly in the absence of a vigorous testing program. There is no realistic threat that a rogue group will be able to make such a lightweight and rocket-deliverable weapon in the near future.

There are several reasons why these small modular reactors may not succeed in the US marketplace. Ironically, one of these is the nuclear licensing procedure. For example, all reactors built in the United States are required to have an "emergency core cooling system." These small modular reactors don't have such systems because they don't need them; they have a much better safety mechanism based on natural convection. Any attempt to add the legally required emergency cooling system—for example, by introducing piping—would interfere with this intrinsic safety and reduce security. But that's the way the rules were written. So the nuclear regulatory commission has to change its rules before these reactors become practical. And changing rules, particularly after widely publicized events like the Fukushima accident, can be politically treacherous for your reelection.

In the United States, the biggest challenge to any new nuclear reactor is the low cost of natural gas—something that affects all alternative-energy systems. But even though natural gas produces only half the carbon dioxide of coal, it still produces some; nuclear power produces none (except for small, onetime amounts in the manufacture of the plant and in the mining of the uranium). If there is pressure to make large reductions in carbon dioxide emissions, then encouraging more nuclear reactors may be one of the most cost-effective ways.

Running Out of Uranium?

Reports persist that there is not enough uranium for a significant nuclear future, but such claims are based on a misunderstanding. This issue was addressed in detail in a long and comprehensive analysis by Kenneth S. Deffeyes and Ian D. MacGregor back in 1978,[27] and what they concluded then continues to be true. They show that although we are indeed running out of the best uranium ore, we are not running out of economically recoverable deposits.

Here's why people are worried. The highest-grade uranium ore contains 10,000 parts per million (ppm) of uranium or more. Only about 100,000 tons of such ore is located in recoverable deposits around the world. A typical nuclear reactor burns a ton of uranium to produce 44 million kilowatt-hours of electric energy. World electric use today is about 130 billion kilowatt-hours per year. Using nuclear power to supply all the electricity we currently use would require 130 billion/44 million ≈ 3,000 tons of uranium every year, and the 100,000 tons of reserves would last only 33 years. If we expand nuclear power, we will run out sooner.

Of course, a lot more uranium would be available if we could use a cheaper grade. If we include ore containing only one-tenth as much uranium, 1,000 ppm, then according to the Deffeyes-MacGregor survey the world reserves are 300 times greater. Instead of lasting only 30 years, the uranium would, in principle, last 9,000 years at the current rate of use—or 900 years if the use went up tenfold. Would it cost too much to use such low-grade ore? Let's look at the numbers.

A kilogram of uranium oxide, extracted from high-grade ore, currently costs about $60. That kilogram, enriched and put in a reactor, can produce about 30,000 kilowatt-hours of electricity. That means that the uranium cost for 1 kilowatt-hour is $60/30,000 = $0.002 = 0.2¢. That's just 2% of the value of the electricity (10¢ per kilowatt-hour). Even if the cost went up tenfold, to 2¢ per kilowatt-hour, it would not require a significant increase in the cost of the electricity—but it would increase our uranium supply by a factor of 300.

Bottom line: We are not going to run out of affordable uranium in the foreseeable future.

Fukushima Deaths

The deaths resulting from the Fukushima disaster were discussed in detail in Chapter 1, so here I'll repeat just the salient points. The best estimate for the deaths from the Fukushima radioactive release is about 100, tiny compared to the carnage of the tsunami. Most of these deaths will not be identifiable among the much larger numbers from unknown causes. That is not an "optimistic" prediction; it is simply a calculation based on the same approach used by scientists to estimate 24,000 deaths from Chernobyl.

Perhaps the greatest future danger from the Fukushima accident will come from the unnecessary abandonment of nuclear power, in Japan and around the world, resulting in economic distress from shut-down plants, and to an increased dependence on fossil fuel, with all of its associated dangers, from war to global warming.

Nuclear Waste Storage

The problem of storing nuclear waste is exacerbated by public misunderstanding. The technical problems have been solved; the remaining problems are ones of perception and public education. The situation is not made better by the fact that some politicians exploit the public fear of anything nuclear. In 2011, German Chancellor Angela Merkel decided to shut down nuclear plants in Germany, even though as a scientist she should know that there were no great dangers; cynics believe she took that action to gain a few more votes from the Green Party, enabling her to continue as chancellor a bit longer.

It is widely known that nuclear waste keeps its radioactivity for thousands of years. Plutonium, a component of the waste in the United States (but not in France, where it is extracted) has a half-life of 24,000 years. And even after that time, its radioactivity is reduced by only half. By 48,000 years, the radioactivity is down to only one-quarter. How can we handle such long-lived waste?

In fact, because plutonium has such a long half-life, it is not a huge contributor to the radioactivity of nuclear waste. Just as important, plutonium is very insoluble in water, so even if it were to leak from a waste storage site, little would dissolve in groundwater. The fearsome reputation of plutonium comes from inhalation; plutonium dissolved in water is far less dangerous. The dose needed to cause one cancer if plutonium-contaminated water is drunk is about 0.5 gram—compared to a lethal dose for inhalation of only 0.00008 gram. It is the threat from inhaled plutonium that gives plutonium its dangerous reputation. Botulinum toxin (sold commercially as Botox) is much worse; its LD50 (lethal dose for 50% of people) is 0.000000003 gram if inhaled. The fact that this number is so small is what is scary about botulinium. It is 27,000 times *more* toxic than plutonium.

I'm not surprised if those facts don't ease your concern about plutonium. The hype in the press has been so great (with the oft-repeated but not true statement that plutonium is the most toxic material known to man) that a few words of reassurance may not be convincing. Perhaps the most important fact is that the experts on waste storage, people who spend their lives worrying about this, think plutonium leakage is not a problem. They worry about the rest of the radioactivity.

And what about the rest of the radioactivity? To put it in perspective, let's compare it to the radioactivity that was removed from the ground when the original uranium was mined, shown in Figure III.12. When the reactor is running, the radiation levels inside are very high, over a million times greater than that of the original uranium. Almost immediately after the reactor is shut down, the radioactivity drops to about 7% of its former level. That level is still very high, and that's why fresh nuclear waste is dangerous, and why an uncooled core will heat and melt.

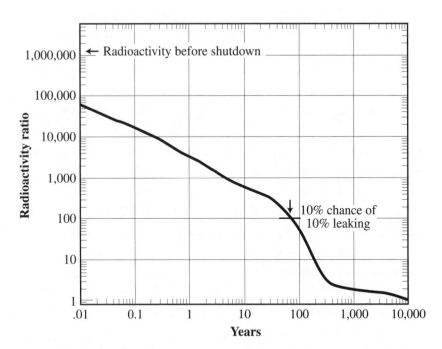

Figure III.12. Radioactivity in nuclear waste compared to radioactivity removed from the ground when uranium is mined. Note that each division is 10x.

Most of the radioactivity comes from atoms that have a very short half-life, but soon those atoms are gone, so the level of radioactivity falls rapidly. After 100 years, the radioactivity has dropped dramatically, as shown in Figure III.12. It is still 100 times more radioactive than the original uranium that was removed from the ground, but it is *only* 100 times more radioactive, and you don't have to store it unprotected in the watershed—the location of the original uranium. If you could build a storage system that had only a 10% chance of leaking 10% of its waste, you'd be back where you started, in terms of radioactivity in the ground. That doesn't sound hard to do, and it isn't. Nuclear waste storage is not a difficult technical problem.

Why, then, are people so concerned about nuclear waste? From my experience, there are three key reasons. First, most people consider radioactivity an unknown and invisible threat, so it is more frightening than are familiar threats like fires and automobile crashes and war. Second, people don't recognize that they are surrounded by a level of natural radioactivity that is usually much higher than the dose that comes from a nuclear accident. As discussed in Chapter 1, the radioactivity near the Fukushima accident dropped within months to a level lower than that normally experienced in Denver, Colorado. Third, the threat of plutonium has been so hyped that many people consider its presence to be unacceptable at any level.

Finally, it just sounds bad. Waste! The premier nuclear waste repository was being built under Yucca Mountain in Nevada. Senator Harry Reid ran for reelection on an anti–nuclear waste agenda: he does not want Nevada to become a *waste dump*! Reid's campaigning may have amplified the Nevadans' perception that the rest of the United States thinks of them as a rogue state (gambling! prostitution!) and a suitable place to dump garbage. It has proven to be effective politics for Reid, and to get Reid's support, presidential candidate Barack Obama pledged in 2008 to shut down the nearly finished Yucca Mountain waste depository in Nevada if he were elected—an action he eventually did take.

In summary, nuclear waste is not a difficult technical challenge. Because of exaggerated public fear, the constraints on nuclear waste far exceed the actual threat, helping to make nuclear power less competitive in the United States. Nuclear power has some advantages over the competition: solar and wind are intermittent; coal and natural gas emit carbon dioxide; practical geothermal is limited in location. The cost of nuclear power has dropped dramatically in the last few years, thanks to reduced downtime, and the new designs can lower the cost even further. The medium-sized modular plants can be placed in remote and difficult locations, offering additional important advantages for the developing world. And nuclear is a technology that China can use to replace its very dirty coal plants.

US development of nuclear power is one of the best ways for the United States to set an example that China, India, and the rest of the world can follow. Please keep that in mind when you are president.

The Coming Nuclear Explosion

In the United States, 31 states have operating nuclear power plants, and in seven of those states, nuclear power supplies more than 50% of the electricity. Interest in nuclear has been undergoing a renaissance in the United States—or at least it was, until Fukushima. Now there is revived public resistance. But regardless of whether the United States develops new nuclear plants, the rest of the world is forging ahead. As of early 2012, 31 countries ran 443 operational nuclear power stations. China is currently building 27 new plants, and it has 50 in the planning phase, and another 110 proposed. Japan is in the process of canceling nuclear power in its own country (thanks to Fukushima), but it is hoping to manufacture plants to be installed around the world. Currently in France, nuclear power accounts for over 75% of the country's electricity needs, and it exports nuclear-powered electricity to some of its neighbors (including Germany and the United Kingdom) that have restrictions on their own nuclear power.

In June 2011, well after Fukushima, France announced it would invest $1.4 billion in third- and fourth-generation nuclear stations such as small modular plants. French president Nicolas Sarkozy said there is "no alternative to nuclear energy today." (The same was true 30 years ago when the French began their massive nuclear program. They wryly explained that France has "no coal . . . no oil . . . no choice.") Their next plant is due to come online in 2018, although it may be delayed. Also in 2011, the UK government confirmed a list of eight locations for possible new nuclear plants that could be built by 2025.

For much of the world, the advantage of nuclear power can be great. The modular reactors don't require much attention; some of them are designed to run for 30 years before they need to be dug up and refilled.

Uranium fuel, because it is so concentrated, is much easier to transport. Much of China's highway and rail infrastructure is currently used to transport coal from the mines to the population centers. In fact, because of the difficulty of moving coal from inland where it is produced to the coastal areas where it is most needed, China—one of the most coal-wealthy countries in the world—has been importing coal from Australia, whose ships can easily reach the Chinese coast. Even in the United States, almost half of our rail shipments (by weight) are for coal. Shipping uranium is much easier. For the energy produced by one ton of natural uranium, you would have to ship 20,000 tons of coal. Shipping uranium enriched to 19.9% is even easier, by a factor of 28 (the ratio of 19.9% to 0.7%, the level of U-235 in uranium ore). That means you can avoid shipping 560,000 tons of coal by shipping one ton of enriched uranium fuel.

12

FUSION

THE JOKE is that nuclear fusion is the energy source of the future, and will *always* be the energy source of the future. Scientists and engineers in the fusion field hate that joke. They don't think it's funny.

Unfortunately, the joke has a solid historical foundation in the overly optimistic predictions made by people working in the field. Back in 1955, the great nuclear physicist from India, Homi Bhabha, said at an international conference, "I venture to predict that a method will be found for liberating fusion energy in a controlled manner within the next two decades. When that happens the energy problems of the world will truly have been solved forever."[28] That prediction was made nearly six decades ago.

Current predictions for fusion power say that it should be ready in the next two decades. Will it ever catch up with the predictions? I think it will, and likely in this century. Fusion may indeed turn out to be a major commercial power source within our lifetimes, even if its development takes too long to help with our immediate energy security and global-warming needs.

Why the history of optimism? *Fusion* is the process that powers the sun, and proponents love to point this out—it makes fusion seem natural. In the sun, the fuel is ordinary hydrogen, the most abundant element in the universe. (That's why stars keep burning for billions of years.) Hydrogen is also the most abundant element in your body, if you go by the number of atoms rather than by weight. Likewise, hydrogen is the most abundant element in the oceans, again by number of atoms. We will never run out of this fuel—at least, not for millions of years.[29]

The abundance of hydrogen was once seen as a major advantage for fusion over fission power, until people recognized that we won't run out of uranium for many hundreds of years either. And, of course, we will not run out of renewables, at least as long as the sun burns (current estimate: another 5 billion years or more).

Another advantage of nuclear fusion is that it is relatively radioactively *clean*. It produces little in the way of dangerous waste. But it's not completely clean. The hydrogen fusion reaction that is most readily used—and is planned to be used for all the large upcoming reactors—is this:

heavy hydrogen + extra heavy hydrogen → helium + neutron

or equivalently, using the common language of nuclear physics:

deuterium + tritium → helium + neutron

Deuterium is abundant in water, but tritium is very rare. The total amount of tritium in all the Earth's oceans amounts to only 16 pounds. For the initial tritium, the supply will be manufactured in nuclear reactors, but eventually the tritium can be created, "bred," in the fusion reactor itself by sending the neutron from the reaction into lithium.

The helium produced in fusion is not dangerous or radioactive; it is identical to the gas in toy balloons. But the emitted neutrons, while also essential for breeding tritium, cause problems. Neutrons are absorbed by most materials, and when that happens they typically make those materials radioactive. The amount of radioactivity is small, especially when compared to the radioactivity in a uranium fission plant, but the very fact that some is produced is used by opponents to argue against this approach. That fear is partially based on the uneducated belief that all radioactivity is bad and that we could and should eliminate it from our environment.

Some alternative fusion reactions don't produce the troublesome neutrons. Most interesting is the hydrogen/boron process:

hydrogen + boron $\rightarrow$ 3 heliums + gamma rays

Gamma rays don't create significant additional radioactivity; they just carry energy. So this reaction seems relatively clean, at least to those who fear radioactivity. Unfortunately, this reaction is much more difficult to set off; it requires higher energy to trigger, and for that reason it will probably not be the first one to be used. In the next section I'll talk about a company called Tri Alpha Energy, named after the three heliums produced (helium nuclei are also called "alpha particles"), that is working hard on this approach.

Finally, we are optimistic because humans *already* produced fusion on Earth, in the form of the hydrogen bomb, way back in 1953. All we have to do is figure out a way to control this fusion, to make it come out slowly instead of in one big boom (although some people have argued that a series of hydrogen bombs underground would boil water to drive turbines, doing just what we need).

There are many different proposals for methods to produce controlled fusion for electric power. I'll discuss five that get a lot of attention and are illustrative of the approaches. They are the tokamak, the National Ignition Facility (NIF), beam fusion, muon fusion, and cold fusion.

Tokamak

The *tokamak* was invented in the Soviet Union in the 1950s; its name comes from an acronym in Russian for "toroidal chamber with magnetic coils." It was soon seen as superior to most of its competitors, and over the past 60 years it has received more attention and more research effort than any other approach to controlled fusion.

The tokamak is based on fusion at extremely high temperatures, also called *thermonuclear* fusion, the same process that takes place in the core of the sun and in the hydrogen bomb. Fusion occurs when two hydrogen nuclei can be brought to touch; the strong but short-range nuclear force then fuses these two nuclei and releases energy. The problem is that both hydrogen nuclei have positive charge, and that makes them strongly repel each other. In *thermonuclear* fusion, that repulsion is overcome by raising the temperature. High-temperature atoms are moving at very high speed; if the speed is sufficient to overcome the repulsion, then fusion can and will take place. In the sun, the temperature of the core is calculated to be about 15 million degrees Celsius.[30]

Fusion in a tokamak requires a temperature above 100 million degrees Celsius. That's 7 times greater than the temperature in the core of the Sun! These superhigh temperatures are needed because we are impatient; we want energy produced very rapidly. The rate of energy produced in the sun is surprisingly small—only about 0.3 watt per liter in the hottest part of the core. That's less than the density of energy release in your own body (75 watts average heat for 75 liters of body averages 1 watt per liter). The sun makes up for this low rate by being large, and all the heat eventually diffuses to the surface. The tokamak can't be large, so it has to increase the rate of fusion by using higher temperature, and also by using a fuel that is more easily ignited, consisting not of ordinary hydrogen but of deuterium and tritium. These forms of hydrogen have extra neutrons in their nuclei; the neutrons don't affect the repulsion (they have no electric charge), but they do increase the nuclear force, and that increases the rate at which the nuclei fuse.

Nothing remains solid at the multimillion-degree temperature of the tokamak, so how can the hydrogen be held? The answer is: in a very fancy bottle made not of material but of a magnetic field. It's

called *magnetic confinement*. The geometry for doing this is not obvious, and most prior attempts led to magnetic bottles that were too
leaky. The tokamak design leaks too, but so slowly that the hydrogen
remains long enough to sustain a thermonuclear reaction, we hope.

Success has been steady but slow. At every stage of higher density
and higher temperature, new leakage issues had to be addressed. As
the systems grew larger and larger, however, the problems began to
subside; size turned out to be a help. The latest version is called
ITER, which stands for "International Tokamak Experimental Reactor" (Figure III.13). (It was originally "International Thermonuclear
Experimental Reactor," but the word *nuclear* scares too many people.) In Latin, the word *iter* means "the way." ITER's goal is to pro-

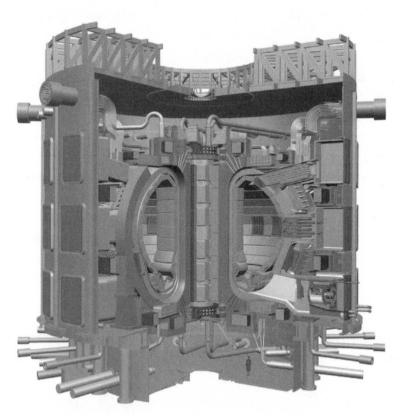

*Figure III.13. ITER, the International Tokamak Experimental
Reactor. For scale, note the figure of the man at the bottom.*

duce 500 megawatts for 400 seconds or more; that's 10 times more power than is needed to run it.

ITER is huge—60 feet (18 meters) tall—and expensive. And recently it has run into cost overruns. The original construction cost was estimated to be 5 billion euros; by 2009, that estimate had risen to 10 billion euros; a year later it was upped to 15 billion. The increase scared more people than did the absolute cost. If the test reactor is so expensive, will fusion ever be competitive? Scientists offered to cut some corners in the research program—for example, by reducing the testing of radiation damage, but such cuts deeply worried other scientists, who feared that reduced testing could lead to greater future problems. If radiation-damaged walls have to be frequently replaced, a tokamak might be hopelessly expensive.

In the official schedule, ITER will inject its first test of hot gases in 2019, start running hydrogen fuel in 2026, and finally reach project completion in 2038. If the results are convincing, commercial reactors might then follow.

There's a generic problem that energy technologies often suffer. When they are distant and abstract, they often seem very attractive to environmentalists. Fusion was originally seen as a cleaner alternative to uranium fission. But when a new technology approaches reality, it is sometimes reexamined and rejected. Greenpeace recently decided to oppose ITER, nominally because of the expense. The organization argues that enormously more value would be obtained if the 15 billion euros were spent on the development of renewable energy technologies. Greenpeace says (correctly) that even if the ITER machine works, it will be many decades before commercial machines can contribute to the world energy needs. They worry that global warming cannot be stemmed by a machine that won't contribute in the near future, and that the money should be spent on wind power, solar, and other renewables that could have an imminent future.

NIF, the National Ignition Facility

In my estimation, the first approach to controlled fusion that will reach *breakeven* (output energy matches input) will not be a toka-mak, but a totally different device based on tiny hydrogen bombs.

The concept is ingenious. Suppose you could make such bombs so small that each produced only 100 megajoules of energy, roughly the equivalent of 5 pounds of gasoline. Exploding 10 of these every second adds up to 1,000 megajoules per second; that's a power of 1,000 megawatts, 1 gigawatt. The energy bursts could be absorbed by a steel tank (a diameter of 10 meters will do just fine), turned into heat, and then used to run a turbine.

The problem is that a conventional hydrogen bomb requires a fission bomb to set it off. Fission bombs require a critical mass of ura-nium or plutonium, and once you have that, you release the energy of kilotons of TNT—much more than the 5 pounds of gasoline equiva-lent you get from the fusion and too large to contain easily. But you don't really need high *energy* to ignite fusion; what you need is high *temperature*. That means a high energy density—and that could be a small amount of energy in a very small volume—and delivered in a very short period of time, so that the heat doesn't have time to radi-ate away. The solution is to use lasers instead of a fission bomb for the trigger. Lasers can deliver energy very quickly, in a billionth of a second or less (light travels one foot in one billionth of a second). If you focus that light on a small spot, it can heat the hydrogen target tens of millions of degrees. A large facility has been constructed in Livermore, California, to accomplish just this. It is diffidently called the "National Ignition Facility" or NIF (pronounced "nif"). The word *ignition* refers to the fact that the lasers supply only enough energy to get the fusion started; after that, the energy released is suf-ficient to trigger the fusion of the rest of the deuterium and tritium. Nuclear physicists think of the process as igniting a nuclear burn.

This approach is sometimes called *inertial confinement fusion*, a confusing term for non-experts. The name derives from the fact that the heating happens so fast that inertia of the fuel keeps it together long enough for the reaction to complete. A big hydrogen bomb also depends on inertial confinement. Contrast that with the "magnetic confinement fusion" of a tokamak. In the tokamak the fusion is continual. In the NIF it will take place in a millionth of a second, with a new fusion burst in a new minibomb every tenth of a second later.

The lasers at NIF are spread out in a building the size of a football stadium, shown in Figure III.14. There are 192 lasers synched together; they take up most of the room. The hydrogen target chamber is in the taller building to the lower right of the image. The laser energy arrives in such a short pulse, often in less than a billionth of a second, that the instantaneous power (energy divided by time) is 500 terawatts—that is, 500,000 gigawatts, the equivalent of a half million large nuclear power plants. Of course, that power is delivered for less than a billionth of a second.

The target is a bit more sophisticated than I've described up to

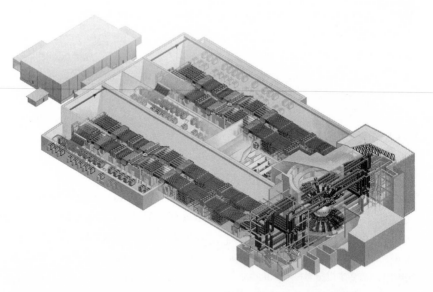

Figure III.14. The National Ignition Facility in Livermore, California. The building contains 192 of the most powerful lasers in the world, and a target chamber (lower right) in which the laser energy is focused to achieve controlled thermonuclear fusion.

now. It absorbs the laser light and heats the outer shell of the target to such a high temperature that the shell emits X-rays; these X-rays not only heat the hydrogen target but create a shock wave that compresses it. The NIF scientists hope to achieve fusion in this manner, but they also expect that in the future they will be able to simplify the target and use the laser power directly to ignite the hydrogen. This improvement is not necessary for producing fusion power, but it is necessary to produce cheap targets that will allow cheap fusion power—power that will be competitive with coal, natural gas, and ordinary nuclear fission reactors.

Skeptics point out that the idea of laser fusion has been "imminent" since the fundamental implosion approach was invented in 1972.[31] As with the tokamak, it took many small steps over many decades—smaller laser systems that tested the principles—to reach NIF. They had vivid names: Antares, Shiva, Nova, Super Nova. The implosion principle was also tested by using nuclear weapons to create X-rays that heated tiny targets; this option ended in 1988 when the United States stopped nuclear testing.

Will NIF achieve breakeven fusion? Yes; in fact, I expect it to happen soon, perhaps even before you read this book. Will it be a useful energy source? Maybe. That depends on whether the cost can be lowered. In my mind, the critical number is the cost of the hydrogen targets, which must be below $1 each.[32] The Livermore scientists have plans for reducing the size and cost of all the components in order to make the package commercially competitive. They call this advanced system *LIFE*, an acronym standing for "Laser Inertial Fusion Energy."

Beam Fusion

Thermonuclear fusion overcomes the electric repulsion of the two hydrogen nuclei by making them so hot that the kinetic energy (energy of motion) of each nucleus is higher than the energy of repulsion. There's another way to overcome that repulsion that doesn't involve heat: take one kind of hydrogen, typically deuterium, and accelerate it into a beam—a plasma of rapidly moving particles. This sounds very much like heating the nuclei, but it is quite different in practice. The beam is easily confined because all the particles are simply moving together; the trick is to have enough of them (beams are usually not very dense) collide with the target.

This approach—called *beam fusion*—is already commercially used. Devices called *neutron generators* accelerate beams of deuterons (the nuclei of heavy hydrogen) in electric fields, and then run them into targets rich in tritium. The fusion of deuterium and tritium produces helium and energetic neutrons, and in these devices it's the neutrons that are valuable. Such machines are, for example, lowered into partially drilled oil wells along with neutron detection instruments. When the device is at the interesting depth, the accelerating electricity is turned on (it takes typically 80,000 volts), neutrons are produced (from fusion!), and they leave the device and enter the rock. Some of the neutrons are scattered back, and the amount of this scattering can give the oil well drilling team a great deal of useful information about the solidity of the rock and its content of oil. The process is one of several methods used for investigating the nature of rock near the drill hole; the entire process is called "oil well logging."

Such neutron generators are valuable whenever there's a need for a lot of neutrons that can be switched on and off. Applications include not only petroleum exploration, but coal analysis in factories, cement process control, measurement of wall thickness, and a host of research and medical uses.

Why not use beam fusion for energy? The fundamental problem is one of efficiency; too few of the nuclei fuse, so the energy output is less than the energy input. Breakeven is not achieved. At least, that's been the case for beam fusion up to now. But clever engineers and physicists keep inventing new ways to improve that efficiency. Perhaps better than pure beam fusion is an approach that combines some of that technology with that of magnetic confinement and more. Tri Alpha Energy is a commercial company in southern California currently operating in "stealth" mode; that is, it is keeping most of its work secret. But Tri Alpha is much larger in scale than most stealth companies; it has raised nearly $100 million to develop its technology. I don't know Tri Alpha's secrets, but members of its research team have given some talks at fusion conferences, and their ideas are intriguing.

The company is hoping to avoid the deuterium and tritium reaction that creates high energy and troublesome (damage-inducing) neutrons. Instead, Tri Alpha's plan is to fuse the nucleus of hydrogen (a proton) with that of boron—the alternative fusion reaction that I described at the beginning of this chapter. The reaction is termed *aneutronic* because it produces no neutrons.[33] In deuterium-tritium fusion the produced neutrons are useful to breed more tritium, but no tritium is needed for hydrogen-boron fusion. The three alpha particles (helium nuclei) produced in the reaction carry off most of the energy.

In Tri Alpha's design, hydrogen and boron are accelerated into smoke ring–shaped beams. The design exploits a rather subtle feature of such rings: they contain both strong magnetic and electric fields. The magnetic fields presumably confine the boron and hydrogen atoms, and the electric fields confine electrons that keep the plasma (charged gas) electrically neutral. The fusion-produced alpha particles, unlike neutrons (which carry the energy in deuterium-tritium fusion), have electric charge, so the energy can in principle be extracted directly into electricity rather than into heat. The trick, as always, is to do this in a way that is cost-competitive.

Will this work? I don't know. It's a big and expensive project involving a good number of top scientists. Don't be surprised if, despite the company name, Tri Alpha first tries deuterium and tritium, since they are easier to fuse. The complexity and subtlety of its approach reminds me of the *magnetron*, a device that made radar possible during World War II—a device so subtle that many physicists today still don't really understand its operation—yet magnetrons have become so pervasive that we all have them in our microwave ovens. Regardless of the success of Tri Alpha, its innovations suggest

that there may be other ideas in fusion waiting to be invented—
ideas that we will take for granted later in this century, just as we
now take for granted our microwave ovens.

Muon Fusion

There is yet another way to achieve fusion—one that doesn't involve
high heat, beams, or electric acceleration. The method was discovered by Luis Alvarez, by accident in 1956. It's original name was
muon-catalyzed fusion, but because it happened at a liquid hydrogen
temperature, –423°F, it was also called "cold fusion." Even though
this discovery was responsible for the name, don't confuse this with
the cold-fusion fad that made headlines in the 1980s; I'll get to that
in the next section.

Alvarez discovered the reaction in his liquid hydrogen bubble
chamber, the device that he used for particle physics research, and
that led to his Nobel Prize. At first what he saw was completely mysterious, but it was soon explained by Edward Teller, and a complete
theory came out a few years later, devised by Berkeley Professor J.
David Jackson.

A *muon* is a small, heavy particle that lives for about 2 millionths
of a second before it explodes, releasing an electron (or positron)
and neutrinos. The key features that make muon fusion possible are
its high mass and the fact that some muons have negative charge—
that is, opposite to that of the proton. Put a negative muon in a
bowl of hydrogen and it will be attracted to a proton, the nucleus of
a hydrogen atom. Because the muon is 207 times heavier than an
electron, it works its way in much closer to the nucleus than an electron can.[34] Now you have a negative muon close to a positive proton
and their electric charges cancel. The electron that was formerly in
orbit around the proton now feels no attraction, and it flies away.

The tiny neutral combination of muon and proton drift freely
through the remaining hydrogen until, by chance, they bump into a
deuteron—the nucleus of heavy hydrogen. It doesn't take long, only

a few billionths of a second, even in ordinary hydrogen; if the material is pure deuterium (as would be used in a commercial reactor) then it happens even faster. When they are close, the strong nuclear force dominates and the proton and deuteron fuse to release energy and to make a helium nucleus.

It doesn't stop there. Most of the time the muon is ejected from the helium; it is attracted to another nucleus, where it cancels the positive charge, and "catalyzes" yet another nucleus. To make a commercial system, the eventual number of such fusions must be great enough that the energy released is more than it took to create the muon in the first place.

The process typically ends when the muon gets stuck to the helium nucleus.[35] Such sticking happens about 1% of the time; unfortunately, that's too often. A simple calculation shows that muon fusion can't work unless you can get each muon to catalyze 350 fusions.[36] So it doesn't work—but we miss by a factor of only 3.5. Can't we find some engineering or physics trick to get us there?

Some people haven't given up. Perhaps it would work if you increased the pressure, or used solids instead of liquids. Some experimenters have indeed managed to eke out 150 fusions per muon. That still isn't enough, but it is a tantalizing achievement.

Can you do better? With a trillion dollars in profit if you succeed, it is worth thinking about. Maybe you can reduce the probability of the muon sticking; an old idea called the "Mossbauer effect," in which a decaying nucleus is put into a crystal, may be relevant, but nobody has figured out how to use it for this purpose. Maybe you could recover the energy of the muon when it explodes so that it isn't wasted; the problem is that the neutrons produced are very effective at escaping the entire chamber. Maybe there's a way to make pions—the precursors to muons—more efficiently, without having to use the heat of fusion to first create electric energy. Maybe you could produce exclusively the kind of pions that you really need—negatively charged ones, without wasting energy making the useless positively charged and neutral ones. That could get you more fusions by a factor of 3—almost enough. Maybe if you added some uranium, the additional energy you would get from the induced fission would be enough. Of course, then it isn't really pure fusion, but it still might be better than a nuclear fission reactor.

It's not crazy; it's just difficult. Right now, it is so difficult that I don't know a solution, but it is conceivable that one exists. Indeed, at least one company, Star Scientific, claims to have solved the problem (or be about to)—maybe using one or more of these the approaches I mentioned, or perhaps doing it some other way. The company says it has developed a way to produce muons that takes less energy. But it has been doing "final testing" for some time now, so it is unclear whether its secret approach really is working.

There may be other approaches to catalyzing fusion. In the 1980s I worked with Luis Alvarez (the discover of muon fission) on a hypothesis he had that there might exist a heavy particle that had negative charge and that did not decay. I set up an experiment to search for such a particle; it would be found in nature, hidden by the fact that there were very few of them. But then I realized that discovery of such a particle could open up a means for fusion catalysis; it could be used in place of the muon to neutralize the charge of the hydrogen nucleus! Any stable, heavy, neutral particle could do this. Alas, we did not find our particle. The theory was wrong. We wrote a nice paper showing that such particles do not exist. But this example shows that there may be something lurking out there that will make simple fusion of heavy water work easily. Keep thinking.

When muon fusion was first discovered, it was called cold fusion because it worked at liquid hydrogen temperatures. But these days the term *cold fusion* is more often used in reference to a crazy form of fusion that uses heavy water and palladium electrodes—one that doesn't work, even though there have been many false scientific claims of success. Some would call this a controversial field, since the proponents haven't backed down, but as a future president, you need to know why cold fusion would be a bad investment for the country.

Cold Fusion

In 1989, two chemists announced a fantastic discovery that looked like it would revolutionize the world's supply of energy. Unlike thermonuclear fusion (but like muon fusion), it could operate at room temperature, so they called it *cold fusion*. It was elegantly simple.

The scientists were Stanley Pons and Martin Fleischmann, researchers in the field of batteries and fuel cells (that is, in electrochemistry) at the University of Utah. They set up a system in which they drove electricity from palladium electrodes through heavy water, to convert the water into its constituent gases deuterium (heavy hydrogen) and oxygen. In some of their experiments, the temperature of the heavy water rose, from the normal 30°C to 50°C. It appeared that the energy coming out was more than they were putting in. Where was the energy coming from? Was fusion occurring? The results were repeatable, but eventually the cell would die and no longer produce energy. Pons and Fleischmann believed that the energy was coming from the fusion of deuterium nuclei. Somehow the palladium and the electricity were overcoming the electric repulsion of the nuclei. Maybe it was a subtle cancellation of the electric forces, as happens with muon fusion.

Calculations showed that such fusion should not happen. There was no plausible way that palladium could overcome the electric repulsion to catalyze fusion. But the greatest discoveries in the past have involved something that was previously thought to be impossible. Experiment typically precedes theory.

Unfortunately, in this case the discovery turned out to be wrong. The mistaken results were a consequence of sloppy experimentation, absence of careful calibration, and the blinding enthusiasm that comes from a belief that a discovery would earn a Nobel Prize, save millions (if not billions) of lives through affordable energy, garner billions (if not trillions) of dollars in royalties, and be remembered as one of the greatest feats of science in the history of humanity.

My personal experience with the Pons and Fleischmann work

might help clarify what happened, and what to watch out for when you are president.

PONS and Fleischmann were using palladium electrodes to electrolyze heavy water[37] when they observed the 20°C temperature rise. But excess heat could be caused by many things, such as a chemical reaction. A key check would be to see whether neutrons were being produced. They looked and found some! Then they searched for tritium, another signature expected in true deuterium-deuterium fusion, and they convinced themselves that they had found that too. They held a press conference announcing their results, and made headlines around the world. Immediately afterward, several teams around the United States set up similar experiments and confirmed the cold-fusion discovery. Prestigious groups at Texas A&M, Georgia Tech, and Stanford all reported verification.

I was as excited as anyone else, and I pondered whether I should change my field of research. I dismissed the skeptics who based their disbelief on theory. I coined a new aphorism: *If it happens, it must be possible*. I was eager to see the paper that Pons and Fleischmann had written describing their work, but it was being kept secret on the advice of the University of Utah patent attorney. Stealth mode. But they did send the paper to several people, and those people sent it to others, and like a chain reaction, suddenly it became available. A friend who had a copy faxed it to me; it was the first fax I ever received.[38]

I read the paper and was stunned—at its low quality. I was profoundly disappointed. The experimental procedures were hardly described; the authors did not talk about calibration of their energy production—an absolutely critical step; they did not describe how they had eliminated other possible energy mechanisms, such as chemical reactions. It was sloppy work—very sloppy work. In just a few minutes, my opinion and optimism turned 180 degrees.

I was very aware of the fact that surprising people make great discoveries that are totally unexpected. Three years earlier someone with the same last name as mine had announced an incredible discovery of high-temperature superconductivity, and that discovery had opened a whole new field of research and many applications. Indeed, back in 1895 when Wilhelm Roentgen discovered X-rays and produced images of his X-rayed hand, for over a week (but only a week) other scientists thought he was a fraud; what he was seeing was clearly impossible in the realm of the currently known physics. But they quickly verified his results. Roentgen received the first Nobel Prize in Physics for his discovery of X-rays.

However, I could not think of a single surprising result in physics (and there have been many) for which the discovery was announced with a bad paper. The Pons and Fleischmann paper wasn't just poorly written; it indicated no sense of self-doubt. In fact, such self-doubt is the essence of the scientific method. Scientists are just as capable of fooling themselves as anyone else, so to compensate, we have to be very careful and skeptical of even our own results. I told my colleagues that if collaborators of mine had written something this badly, it would lower my opinion of them so much that I would no longer collaborate with them.

Rather than starting my own research program on cold fusion, I started taking bets against it. I offered 50-to-1 odds that the Pons and Fleischmann discovery would not hold up. My betting was reported in the *Wall Street Journal*, but not my "it was a bad paper" logic for being willing to give such odds.

What about the teams that reported confirmation of the Pons and Fleischmann results? In retrospect, there appears to have been a case of publication bias. If you do an experiment and get a fluctuation (or maybe make a mistake) that seems to verify the result you're seeking to test, you might be tempted to quickly write a paper and maybe even call a press conference. After all, someone else has already reported the claim; it would be very strange for your experiment to verify that claim if it weren't real. So maybe you won't be as careful as you normally are. On the other hand, if you see nothing, you don't write a paper because you assume you might have done something wrong. And besides, negative papers are rarely accepted for publication in journals.

Soon after publishing the corroborating papers, though, many of the teams that had "verified" the results changed their conclusions and withdrew their claims. This happens in science when you check and recheck everything you did and have colleagues help in the checking. In your initial rush to publication, you may not have examined all other sources of energy that could have heated the water. Maybe you made some mistakes with your gamma calibration. Maybe the tritium you saw was from contamination. (Many people who know how to detect tritium already have some in their labs, for other projects.)

In addition, Pons and Fleischmann were caught in some blatant mistakes. The data they published on the gamma rays did not match the data you would get from fusion; the energies were wrong. They could not explain the discrepancy. A year later, Pons and Fleischmann published a detailed paper on the heat production, but this time they didn't even mention measurements of gamma rays. Why not? Did they try but fail to verify their prior claims? One of the key aspects of the scientific method is candor. If you find that some prior results you got were wrong, in science (unlike politics) your colleagues expect you to explain what went wrong, openly and publicly. This team did not do that.

Finally a session at a meeting of the American Physical Society, the premier physics organization, was arranged to evaluate the entire story. Nine reviewers presented reports, including what they had learned from visits to the Pons and Fleischmann laboratory. One of the speakers told me that he could find no evidence that Pons and Fleischmann had ever calibrated any of their instruments. Steve Koonin, then a professor at the California Institute of Technology (later to become an undersecretary at the Department of Energy), declared at the meeting that the claimed discovery was due to "the incompetence and delusion of Pons and Fleischmann."

Although the cold-fusion discovery had been discredited in the scientific community, the University of Utah hoped it was still valid (the patent rights could make it into the best endowed university in the world), so it gave Pons and Fleischmann $4.5 million for continued research. Since their initial budget had been only $100,000, that represented a boost by a factor of 45, surely enough to address all the complaints of critics. They set up a new laboratory—but after a few years of disappointing results, the lab funding was ended. Pons eventually received $40 million for additional work from Toyota—400 times his original budget—but that program, too, was eventually terminated. Cold fusion did not disappear because of lack of money.

Claims of cold fusion persist. In 2006 I met a representative of the Space and Naval Warfare Systems Center in San Diego, California. When I told him my negative opinion of cold fusion, he told me I was completely mistaken and said they were doing research that indicated there really was something there. He gave me a coy smile that reminded me of the Mona Lisa. In April 2009, the CBS program *60 Minutes* announced that cold fusion was again "hot."[39] The show featured programs at SRI International in California and at Energetics Technologies in Israel. More recently, scientists at the University of Bologna claimed to have achieved cold fusion and hope soon to build a device that delivers 10 kilowatts of commercial power. They won't let anyone see how it works because they say they're having trouble getting their patents approved. In 2011, a new company in Berkeley named Brillouin Energy Corporation made fantastic claims about cold-fusion energy production, but a few months later the company was already backing off, and said it was producing only a small increase in energy (a factor of 2 more than it takes to run the system).

Why is the cold-fusion story significant and worth telling? There are several reasons. One is that fantastic claims in the energy field will persist long after they have been discredited. True scientific discoveries (such as X-rays or high-temperature superconductivity) are usually quickly verified. In fact, the scientific community is very open to new discoveries, but it enforces high standards of demonstration. Those standards are the essence of the scientific method.

Yes—support wild ideas. But pay attention. If the quality of the work is poor, then the wild idea is probably wrong. If you cannot examine the work (a frequent excuse is that it must remain proprietary until patents are approved), then be suspicious.

To learn more about mistaken discoveries, I highly recommend an essay by Nobel Laureate Irving Langmuir titled "Pathological Science." It is available online at

http://www.colorado.edu/physics/phys3000/phys3000_fa10/
langmuir.pdf

13

BIOFUELS

IT'S HARD to write about biofuels because they trigger so much passion. There are true believers and skeptics and people everywhere in between. In fact, biofuels are so full of contradictory implications that I suspect this chapter will offend more people than any other in this book, with the possible exception of the chapter on electric cars. Let me list some of the possibly contentious conclusions that I will defend. (Some of these may not be surprising to you, but they will be to many others.)

- Corn ethanol should not qualify as a biofuel. Its use does not reduce greenhouse gas emissions into the atmosphere. Likewise for paper and waste oil from restaurants; they might be called *pseudo-biofuels*, because they offer no global warming advantage over fossil fuels.
- Biodegradable and recycling are overhyped; from a purely global-warming perspective (ignoring the aesthetic aspects), you could even call them *bad*.
- Ethanol from cellulose—the stalks of rapidly growing plants such as switchgrass and *Miscanthus*—offers the best hope for a significant biofuel component to solving our energy problems.

• The main value of biofuels is not in reducing global warming, but in increasing energy security. For that reason, their main competitors will be shale gas, synfuel, and shale oil.

Let's try to examine biofuels objectively, look at the arguments, discover and dismiss the hype, and determine their true value.

Ethanol from Corn

Corn ethanol should not qualify as a biofuel. The reason is that its production requires a great deal of fertilizer, as well as oil and gasoline to run the farm machinery, plow, harvest, and deliver. In the end, it just doesn't produce enough sugar per acre (turned into ethanol by fermentation) to yield net carbon dioxide reduction. This point has been hotly argued, but the best calculations (now generally accepted) have shown this neutrality. On the basis of such conclusions, in 2011, the US Senate voted to eliminate the subsidy of 45¢ per gallon for the combining of ethanol with gasoline to make "gasohol."

Why was ethanol originally labeled as a biofuel? Part of the reason is scientific naïveté. It seemed obvious that anything grown would be carbon-neutral—until all the ancillary carbon sources were added up. Another reason may have been politics. Iowa is a state that gets a particular benefit from the rising prices of corn, and Iowa is also the state with the earliest presidential caucus. A candidate for president takes a great risk by adopting a position that could hurt the income of Iowa. So beware.

Another downside of corn ethanol is its impact on food prices. Tim Searchinger of Princeton wrote in a 2011 article: "Since 2004 biofuels from crops have almost doubled the rate of growth in global demand for grain and sugar and pushed up the yearly growth in demand for vegetable oil by around 40 percent. Even cassava is edging out other crops in Thailand because China uses it to make ethanol."[40] What's good for Iowa is not necessarily good for Mexico, where the price of corn for tacos has risen. All this for no greenhouse effect benefit.

Of course, there is another reason to justify corn ethanol: energy security. Corn ethanol is a domestic product, and it reduces our dependence on imported oil. Corn for ethanol production in the United States grew from 500 million bushels in 1999 to 10 times that—5 billion bushels—in 2011. One bushel can be made into 2.8 gallons of ethanol. So the corn made 14 billion gallons of ethanol. Per gallon, corn ethanol delivers only two-thirds the energy of gasoline, so that is equivalent to 9 billion gallons of gasoline, about 3% of US consumption and about 5% of US imports.

That is a significant reduction, and it does help the balance of trade; money that would otherwise go to the oil cartel instead goes to Iowa farmers. As president, you will need to gauge this benefit against the higher food costs. And you must consider the alternatives. For the purpose of energy security, there are synfuels and natural gas and soon shale oil. But keep in mind that the main public selling point for corn ethanol—its carbon neutrality—is scientific nonsense. If you claim otherwise, your opponents will attack, and they'll be right.

Biodegradable Is Bad?

You're probably not surprised to hear me say that the passion for biodegradable everything is overhyped; but biodegradable is *bad*? That's on the list at the beginning of this chapter, and it goes against decades of teaching in our public schools. But it is bad only from the point of view of greenhouse emissions. The original interest in biodegradable materials arose because of their harm to aesthetics and to animals: plastics enduring and floating in the oceans, strangling birds and sea otters; plastic and Styrofoam cups persisting in the wilderness, destroying our sense of beauty and of nature—the sometimes illusion that we are remote from civilization. When materials containing carbon, such as plastics and oils, biodegrade in the presence of air, they produce carbon dioxide, the primary greenhouse gas. If your bigger concern is not aesthetics but global warming, then it is better to bury these materials, to sequester them, to put them underground in such a way that they won't biodegrade for the foreseeable future.

I confess that my complaint is a bit tongue-in-cheek. Biodegradable isn't really bad; the carbon dioxide emitted is negligible, and as a backpacker, I really care about the aesthetic aspects. But I do get annoyed by people who push biodegradable *everything* with religious fervor, people who want to outlaw everything plastic as if it were intrinsically evil.

Pseudo-biofuels

Suppose you use waste cooking oil for fuel. Isn't that a biofuel? In some sense it is—it was made from plant matter—but so was coal. Using waste oil as fuel (rather than burying it) adds carbon dioxide to the atmosphere; it's no better than petroleum. Running your auto on it is no better for greenhouse gas reduction than is running the car on gasoline. It would be different if you used *primary* vegetable oil, oil made specifically for autos from plants that otherwise would not have been grown. In that case, growing the plants removes from the atmosphere some carbon dioxide that wouldn't otherwise be removed. Waste cooking oil is not biofuel; I refer to it as a pseudo-biofuel because it gives the illusion of being good for the environment. You certainly should not earn a tax credit for using it.

Recycling of paper is also bad for atmospheric greenhouse gases. If you bury it (and don't let it biodegrade), then you are sequestering carbon and someone has to grow a new tree to make the replacement paper, and doing that removes carbon dioxide from the atmosphere. Paper used for a fuel is pseudo-biofuel, as is any waste product that would otherwise be buried. Environmental and governmental organizations should stop bragging that their publications are "made with recycled paper"; using such material results in more carbon dioxide in the atmosphere! Some people are horrified when they discover 50-year-old newspapers still readable in old landfill. From a global-warming perspective, they should celebrate.

I don't want you to get the impression that burying newspapers will make a big dent in the world's atmospheric carbon dioxide. I am only pointing out the conflict in the logic of some pedants who look down on anyone who doesn't recycle.

Rotting garbage in landfills can be converted by anaerobic bacteria into methane gas—but that, too, is pseudo-biofuel. The Altamont landfill in California can produce 13,000 gallons of liquefied natural gas every day; this natural gas is used to operate most of its waste and recycling trucks. Altamont says it captures 93% of the methane produced. That means 7% leaks to the environment. That's unfortunate, since methane is 23 times more potent (meaning bad) as a greenhouse gas than is carbon dioxide. It would have been better (from a greenhouse perspective) to burn the garbage and turn it into carbon dioxide, or even better, to bury it in a way that preserved the garbage as garbage. If leakage can't be reduced, I suspect that concerns about global warming will make this approach unpopular.

My wife Rosemary, an architect, jokes that she should earn carbon credits for building buildings. Carbon (in wood) is sequestered for as long as the buildings last, and even longer if the old wood is buried or reused. *Architectural sequestration!*

Let's now look at the serious biofuels—fuels from plants chosen specifically for that purpose. Can they truly help us reduce carbon dioxide emissions and achieve energy security?

Ethanol from Cellulose

Sugar is sweet; cellulose is strong. Sugar is easy to digest, and it (as well as starch) can be converted by yeast into a good liquid fuel, ethyl alcohol. Cellulose is a long chain of sugar molecules, but you wouldn't know that from its indigestibility, a trait that not only allows it to be the building material of tall trees, but also protects it from most predators. Cows can eat cellulose, but they have to use an elaborate digestive system with four stomachs and work full time at it. Deep inside the cow are protozoa and bacteria that convert the cellulose to "fatty acids" that can be absorbed into the blood and deliver usable energy.

Corn is, technically, a grass that was cultivated for the first time a few thousand years after the end of the last ice age. Its key property, as far as humans are concerned, is its tasty, starchy, and digestible kernels. Though the original kernels were small and tough, humans have cultivated them to be large and sweet. But they still make up only a few percent of the mass of the corn. To make truly efficient use of the plant, we need to make ethanol from the stalks, from the cellulose.

If cows can do it, we should be able to do it too, perhaps by developing microorganisms that convert cellulose directly into ethanol or another fuel. This technology is under rapid development. Unlike cows, industrial processes can operate at high temperatures, and that appears to be a significant advantage. One approach is to identify and perhaps develop fungal enzymes that can break down the cellulose. A fungus discovered in Patagonia, *Gliocladium roseum*, appears to do just this. There is also a kind of yeast that can create ethanol from straw. (I would have named it the Rumpelstiltskin yeast.)

Cellulose conversion is a rapidly developing area of research that takes full advantage of the numerous recent discoveries and developments in molecular biology. I am optimistic that additional solutions will be found, in the near future. It is too early to tell which of these will prove to be commercially viable.

Once you can convert cellulose, then you no longer need sugary corn, and you can instead concentrate on the fastest-growing cellulose producers. Top candidates include *switchgrass* and *Miscanthus*, a grass that grows over 11 feet tall and can yield three crops per year. It is resistant to weather damage and can produce, in principle, 1,150 gallons of ethanol per acre per year, whereas corn (even if you include all the cellulose) can produce only 440.

Can we devote enough cropland to Miscanthus to have an impact? Cellulose contains about one-third the energy of an equal weight of gasoline, so for a ton of oil you need to grow 3 tons of Miscanthus—assuming you can convert it with no energy loss. US oil use—22 million barrels per day—amounts to a billion tons per year. So we would need 3 billion tons of Miscanthus each year. How much can we grow? Ask any farmer, and you'll hear that 5 tons per acre is difficult, but the Miscanthus optimists assume 15 tons per acre. Sticking with this optimism, we'd need 200 million acres. That's a square 560 miles on each side, nearly six times the size of Iowa, about half of the cropland of the United States. And that estimate assumes optimism in the conversion efficiency and in the yield per acre. The numbers aren't good. Miscanthus can supplement our needs, but unless the optimism proves right, and unless we devote huge resources to its production, biofuel is unlikely to provide a substantial portion of our liquid fuel demand.

Ethanol from Algae

Potentially even better than grass is algae. There is hope that, instead of making the intermediate product of sugar or cellulose, the right kind of algae can create oils (lipids) that can be used in diesel engines without any costly conversion step. Algae can use sunlight very efficiently; essentially every cell in the algae actively produces biomass—unlike grasses, which do so only on the leaf surfaces. Proponents claim that algae can produce more than 10 times the energy per acre that Miscanthus can produce. A great deal of work is being done to develop algae for this purpose, much of it in the commercial (versus government-sponsored) world. Companies are inducing mutations and using genetic engineering to modify algae to match their needs. Moreover, algae can be grown in salty or brackish water.

Despite these scientific and commercial advantages, I think it's unlikely that algae will become economically competitive, especially when compared to Miscanthus. Grasses grow in large, open fields, endure tough weather, and can outcompete invasive species.

In contrast, experience indicates that algae optimized for fuel production are extremely susceptible to invasion by rogue algae and bacteria, even if grown in covered containers.

IN THE END, I do not believe that bioethanol or other biofuels will prove important for limiting the greenhouse effect. Biofuel could substitute for gasoline, but doing that for the United States would have only a tiny effect on the predicted world temperature rise, probably less than $1/40$°C.[41] For energy security, bioethanol will come along too late, and be too expensive, to truly compete with compressed natural gas, synfuel, or shale gas.

14

Synfuel and
High-Tech Fossil Fuels

THE UNITED States is running low on oil but it is not running out of natural gas and coal. That's good news for energy security, but bad news for greenhouse emissions. The immediate problem is that our vehicle designs and infrastructure are built around oil. It is possible that we could convert quickly to natural gas, although doing that requires larger fuel tanks or reduced vehicle range or both. As I'll show in detail in Chapter 16, I believe that all-electric vehicles will be too expensive to be truly competitive.

There are fossil fuel alternatives. We already discussed shale gas in some detail in Chapter 4 and shale oil in Chapter 6. Other unconventional sources of fossil fuel include synfuel, manufactured from gas or coal; methane extracted by fracking coal beds; gas created by setting deeply buried coal beds on fire; and oil extracted from places that we once thought were uneconomical but that become exploitable when the price of oil rises.

Synfuel

We are low on liquid fuels for transportation, but we have plenty of coal and natural gas. Why not use these fuels, add a little hydrogen, and use chemistry to manufacture synthetic oil? Sounds simple, and obvious. What's the catch? Cost?

No, not really. In fact, in late 2011 Sasol, a major South African corporation, announced plans to do just that, to manufacture oil from natural gas. Sasol has been combining coal and water to make diesel fuel ever since the apartheid era, when the major oil producers embargoed shipments to its country. The chemistry had first been developed by Germany (Figure III.15), which had used the Fischer-Tropsch process (a tricky but effective chemical method) during World War II to convert its coal to oil. Today, in our acronym era,[42] we call the method GTL (for "gas to liquid") or CTL (for "coal to liquid").

Figure III.15. The ruins of the German World War II synfuel plant in Politz, Poland, where about a third of the Nazis' aviation fuel was produced from coal.

The new Sasol plant will be built in Louisiana to take advantage of nearby natural-gas discoveries. The plant will cost $10 billion and will create 100,000 barrels per day of diesel fuel. At $100 per barrel (the price as I write this), the synfuel is worth $10 million per day, $3.6 billion per year. Subtract the cost of the natural gas and of operations, and Sasol should still get at least $3 billion annual income from its $10 billion investment. That's a 30% annual return.

Why aren't we building many more such plants? We were, back in 1976 after the first OPEC oil embargo. A high rate of return was expected—but then the price of oil plummeted to $22 per barrel (in current dollars). In the future, Saudi Arabia can undercut any threatening technology as long as it has surplus capacity, since it can pump oil for under $3 per barrel. Moreover, the price of natural gas dropped, and many power plants and homes switched from oil to this now cheaper alternative. That, too, undercut the synfuel investment (although it lowers the cost of the feedstock).

The United States tried to get synfuel going again in 2007. With domestic supplies of oil running low, the US military faced a national security threat. You can't fight without fuel; the fear was that OPEC could emasculate our forces by a simple embargo. To protect the United States from this threat, President George W. Bush signed the Energy Independence and Security Act. In its original version, it had provided loan guarantees for up to 10 large coal-to-liquid synfuel plants, each costing $3 billion or more. It had tax credits for synfuels sold through 2020, and automatic subsidies if the price of oil dropped below $40 per barrel. The air force would sign a long-term contract with the synfuel producers, giving them a guaranteed customer. The synfuel section was initially supported by then presidential candidate Barack Obama.

But prior to passage, the synfuel sections were eliminated from the act, largely because opponents pointed out that coal produces about twice the CO_2 of oil. (Coal gets all its energy by burning carbon to CO_2; oil contains both carbon and hydrogen, and the H_2O produced is harmless.) Add in the fuel used to drive the Fischer-Tropsch plants, and the net carbon dioxide produced is even greater.

In 2007, global warming was a major public issue. Many US citizens were deeply embarrassed that the United States had not ratified the Kyoto Protocol limiting greenhouse gas emissions, and they hoped that would change if Barack Obama was elected. So, largely for environmental considerations, the bill was rejected along party lines, with the majority Democrats holding sway. Warming worries had trumped security. But there was a great deal of confusion. From his press releases at the time, I believe that Democratic Senator Barack Obama supported the energy independence bill because he didn't realize that synfuel results in more CO_2 than does fossil fuel, and when used in automobiles, there is no way to sequester this greenhouse gas.

As a future president, you may want to reevaluate this trade-off between warming and national security. Recall that the US automobile has contributed about $\frac{1}{40}$°C to global warming. In the next 50 years, assuming we adopt reasonable automobile emission standards, we should be able to limit the temperature rise attributable to the US automobile to an additional $\frac{1}{40}$°C. A switch to 100% synfuel would boost that to about $\frac{1}{30}$°C. The danger of that much rise is what you need to balance against the possible national security needs. In addition, you might want to consider the role that synfuel might play in reducing the balance-of-payments deficit.

The possible downsides to the Sasol investments include a potential drop in the price of oil (it could happen if the developing world joins us in a deep, long-lasting recession) or an increase in the price of natural gas (unlikely because of the large US and foreign reserves) or a rapid expansion in the exploitation of shale oil, a very realistic possibility. Of course, if the price of gasoline rises, then the Sasol profits will rise too.

I expect a large expansion of synfuel facilities in the United States over the coming decades as their profitability becomes evident, unless such growth is inhibited by legislation. The subsidies and guarantees that the industry originally required are no longer necessary, thanks to the abundance and low price of natural gas. In 2000, Chevron entered into a joint venture with Sasol called

Sasol Chevron Holdings, and they have already begun operations in Qatar. Expect more plants in the United States.

There are other possibilities for synfuel. The Great Plains Synfuels Plant has been operating since 1984 in North Dakota to produce synthetic natural gas (unnatural gas?) from a low-grade coal called lignite. Synfuel is a big subject, and I expect it will play a central role in future US energy policy.

Coal Bed Methane

When coal beds are too deep to be mined (some are half a mile down), they can still be exploited if there's methane gas in the coal bed. Such methane can be extracted by the simple method of drilling down to the coal and letting the gas escape through natural cracks, sometimes after draining the well of water. Fracking with horizontal drilling can be applied to enhance the recovery. Substantial coal bed methane is extracted from mines in Wyoming and Montana; about 7% of the methane currently produced in the United States comes from this technique. Although coal bed methane often contains carbon dioxide and nitrogen, it does not have the rotten-egg gas hydrogen sulfide (H_2S) that is found in conventional natural gas sources. That gives it the nice nickname "sweet gas." It is also relatively free of heavier hydrocarbons such as propane and butane. Because of its purity, it is typically referred to as methane rather than as natural gas. (Shale gas is also relatively pure, with little or no H_2S, nitrogen, or carbon dioxide.)

Coal Bed Gasification

An ingenious method to exploit deeply buried coal was inspired by underground coal fires. A deposit in Australia now called Burning Mountain was ignited by lightning several thousand years ago, we think, and it burns still. A coal mine ignited in 1962 has turned Centralia, Pennsylvania, into a ghost town. But the coal in these deposits does not burn to completion; it typically emits large amounts of hydrogen gas and carbon monoxide. Those are useful fuels.

Such partial burning of coal, in place underground, is now being tried as a means of extracting energy from beds that are too deep to mine directly. You pump down air or oxygen, add some water (or let natural seepage take care of that), and ignite (Figure III.16). The coal partially burns to make carbon monoxide, and some of the resultant heat makes other coal react with water to produce hydrogen and more carbon monoxide. The resulting mixture is called "coal gas." It is remarkable that this *coal bed gasification* can be done

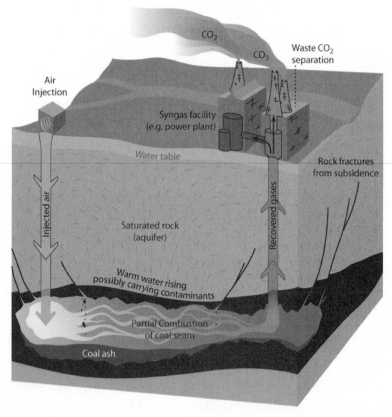

Figure III.16. Burning coal underground—coal bed gasification.

deep underground. It is the ultimate in remote chemistry. The temperature in the burning coal seam can reach 1,500°F.

The main advantage of coal bed gasification is that it allows the energy of the coal (or at least much of it) to be recovered without the coal itself having to be dug up. Moreover, the ash is left in place; there's no need to dispose of it (and coal ash is carcinogenic). The product can be used for more than just energy; coal gas is the starting point for the Fischer-Tropsch process to manufacture synfuel, and it can also be used as the feed gas for other chemicals, including methanol. The disadvantages of underground gasification are that some energy is lost by heat conducted away underground, some coal is left unburned, and there is a danger of polluting the water table.

Will this method become widespread? I don't know. It still seems miraculous to me that such a method actually works.

Enhanced Oil Recovery (EOR)

In the early days of oil wells, the oil came gushing out of the ground. The enormous pressure came from the weight of the rock above the deposit. But if you simply depend on this pressure, you'll typically recover only 20% of the oil in the deposit. Why? Diagrams of oil wells sometimes give the misimpression that oil is found in huge otherwise empty caverns. In fact, oil is found in somewhat porous and cracked rock, with most of the volume occupied by the rock. Some of it flows freely, but there is often a substantial fraction that sticks and does not flow, even under great pressure. In addition, some valuable components of the oil are viscous and don't move easily through the narrow cracks.

You can get more, up to 40% of the oil, by flushing it out with natural gas, water, or carbon dioxide. Such extraction is called secondary oil recovery. There is an added benefit if you use carbon dioxide, since then you're sequestering it—keeping it out of the atmosphere—and that has global-warming benefits (and in some countries will earn you carbon credits). Unfortunately, depleted oil wells will not provide enough volume to sequester more than a small fraction of the carbon dioxide we produce.[43]

The goal of *enhanced oil recovery* (EOR) is to extract some of the remaining 60% of the oil that is still down the hole. This is a potentially enormous resource. The rising price of oil has made it feasible to use methods that were inconceivable just a few decades ago. These include heating the oil to reduce its viscosity by inject‑ing steam; a related method is pumping down air or oxygen to allow in‑situ burning of some of the oil. You can pump down soap (sur‑factant) to wash the oil off the surfaces of the rock. You can send down certain kinds of bacteria, related to the oil‑eating bacteria that consumed so much of the Gulf spill, to break down the more viscous molecules of oil (those consisting of long chains) into short ones that flow more easily. Bacteria have the advantage that if they start growing underground, you don't have to keep adding more; they are, in effect, free. You want, of course, bacteria that feed on only the heaviest, longest molecules, and not on the shorter ones (from propane to octane), which are the most valuable fuels.

Oil Sands

Here's a trivia question: Saudi Arabia and Venezuela have the larg‑est recoverable oil reserves in the world. (Ignore shale oil for the moment.) What country comes in third?

The answer is adanaC. (I spelled it backward so you wouldn't acci‑dentally see the answer before you guessed.) But this country wasn't in third place a decade ago; it took that position only when the price of oil rose above $60 per barrel. And it lost its third‑place position when, briefly in June 2007, the price dropped below that value.

The explanation for that complicated behavior is simple: most of Canada's oil is in the form of very heavy crude oil called bitu‑men, mixed with clay and sand. These are called *oil sands*, or some‑times (but inaccurately) *tar sands*. The reserves are enormous. The International Energy Agency estimates about 200 billion barrels, but one of the oil companies that is exploiting the reserves, Shell Oil, believes they may be 10 times larger, 2 trillion barrels. That would mean that Canada has the greatest oil reserves in the world (again, ignoring shale oil), substantially greater than those of Saudi Arabia. To put these numbers in perspective, with 2 trillion barrels Canada could supply the total US current consumption for over 250 years; it could supply the entire world (assuming no energy growth) for over 60 years.

The Canadian oil sand, like oil shale, demonstrates the nonsense of a naïve application of Hubbert's law, the theory that any reserve will follow a predictable bell-shaped curve. Hubbert's law doesn't work very well if there is a price threshold that, when crossed, makes huge formerly untapped reserves economically recoverable.

There are several objections to exploiting the oil sands. The Canadian deposits are near the surface, where ugly open-pit mining is feasible. The dangers of local water pollution are large, and the nearby boreal forest is being destroyed. How do you balance environmental considerations against the need for oil? Moreover, the mining operation requires large amounts of water. The oil companies say that the amount required is about 20% of the volume of oil produced, but environmentalists have disputed that number, saying it is actually 10 times higher. Of course, the answer depends on how much water is reused.

For the more deeply buried oil sands, the oil is extracted by injecting steam to reduce its viscosity. This process requires a large energy expenditure. The current source is natural gas, although there have been proposals to build a nuclear reactor on-site. In principle, the oil itself could supply the energy. The energy it takes to extract one barrel of oil is about 12% of the energy in the oil, so this is feasible.

The immediate competitor for this Canadian oil is synfuel from natural gas, and the long-term competitor is shale oil. The price of synfuel is about the same: roughly $50–$60 per barrel to produce. This fact makes it unlikely, in my mind, that the long-term price of oil will be sustained at much above this value. And it may go even lower; some industry experts say that shale oil can be recovered for $30 per barrel.

15

ALTERNATIVE ALTERNATIVES: HYDROGEN, GEOTHERMAL, TIDAL, AND WAVE POWER

Hydrogen

WHATEVER happened to the hydrogen economy? In his 2003 State of the Union address, President George Bush announced it as a major initiative. Here are his words as reported in the *Washington Post*:

> Tonight I'm proposing $1.2 billion in research funding so that America can lead the world in developing clean, hydrogen-powered automobiles.
>
> (APPLAUSE)
>
> A simple chemical reaction between hydrogen and oxygen generates energy, which can be used to power a car, producing only water, not exhaust fumes. With a new national commitment, our scientists and engineers will overcome obstacles to taking these cars from laboratory to showroom, so that the first car driven by a child born today could be powered by hydrogen, and pollution-free.

(APPLAUSE)

Join me in this important innovation to make our air significantly cleaner, and our country much less dependent on foreign sources of energy.

What happened to the grandiose vision of President Bush? The answer: nothing, because that vision was fundamentally flawed. Hydrogen automobiles were never a good idea. They have two major failings that they share with battery-driven automobiles.

First, hydrogen is not a source of energy, but only a means of transporting it. We don't mine pure hydrogen or pump it from the ground; we get it from water (H_2O) through electrolysis, or from natural gas (mostly methane, CH_4) by chemical reaction with water. Electrolysis takes energy, and when we use the hydrogen as fuel we get back only part of what we put in. When we get hydrogen from methane, carbon dioxide is still created, so that process does not address global-warming fears. And it is cheaper and more efficient to use the methane itself as a fuel, either by combusting it or by using it in a methane fuel cell.

Second, hydrogen takes up a large volume, so a hydrogen automobile will have either very short range or a huge gas tank or both. Many people are misled by the fact that hydrogen contains 2.6 times the energy of gasoline per pound. That's true, but a pound of hydrogen takes a lot more space. Per volume, per gallon, it isn't very convenient. At maximum pressure, it takes 10 gallons of hydrogen gas to match the energy in 1 gallon of gasoline. That's for an internal combustion engine; but even if you use hydrogen in a fuel cell to drive an electric engine, it still takes 6 gallons of hydrogen to match 1 of gasoline.[44] In principle, liquefying the hydrogen increases the energy per gallon by a factor of 3, but that takes very low temperature (−474°F, very close to absolute zero) and a super thermos to make sure it doesn't boil off. Although hydrogen rises rapidly in air, if trapped in your garage it can explode over an unusually wide

Figure III.17. The Honda FCX Clarity, one of the first hydrogen-powered automobiles.

range of mixtures: from 4% up to 75% hydrogen in air. (Natural gas has a much narrower range: 5%–15%.) These facts also make it very difficult and dangerous to transport and store hydrogen for filling stations.

Despite these negatives, a hydrogen automobile, the Honda FCX Clarity (Figure III.17) was produced in limited numbers starting back in 2008. The Clarity is sold (leased actually) only in Los Angeles, where 10 hydrogen filling stations have been set up by Honda. This rollout was clearly experimental, not only to gauge public interest, but also to learn the problems involved in supplying hydrogen fuel.

Hydrogen fuels are great for rockets, since for them weight is more critical than volume, and elaborate safety measures can be taken to allow the use of liquid hydrogen. Hydrogen might conceivably make sense for airplanes. But for autos, having super lightweight fuel doesn't add value.

Some people argue that hydrogen autos could have a 300-mile range; all we have to do is make the autos lighter and more efficient. That is true, but such improvements would also benefit other fuels. With similar reductions in weight, cars running on gasoline that now get 35 mpg could then get over 100 mpg.

The main appeal of the hydrogen auto is its potential to eliminate carbon dioxide emissions. But doing that requires that we obtain our hydrogen by electrolysis using a low-carbon electric source such as solar, wind, or nuclear. It is worthwhile remembering that the global warming expected in the next 50 years from US automobiles is only about $\frac{1}{40}$°C under reasonable efficiency standards. Moreover, this temperature rise will be even lower if we achieve 100 mpg for gasoline autos—obtainable by using lightweight materials. The big issue driving our move away from gasoline has been transportation energy security, not climate change.

The competition from natural gas will be tough on hydrogen. It is remarkably easy to convert an ordinary gasoline engine to run on natural gas or on both natural gas and gasoline. We'll discuss this further in Chapter 17.

If hydrogen vehicles are so impractical, why do car companies introduce them and announce plans for increased production? I suspect they do it for good green public relations. Automobile companies are constantly under public attack for many reasons—some real, many imagined. Showing that they're working hard on flashy new technologies for the public good is good advertising. But look carefully at their claims. If the number of cars produced is actually very small, then suspect that the company isn't really serious. Media coverage can be great, cheap advertising. In 2009, *Fortune* magazine estimated the cost of producing the hydrogen-fueled Honda Clarity at $300,000 per car. In 2010, Honda announced that 50 of these cars were available for lease in the United States, and its goal was to have 200 worldwide. Honda did say it could have large-scale production as early as 2018 but sales might be limited by the lack of hydrogen refill stations. Don't bet on that large-scale production coming anytime soon.

Geothermal

Like wind, heat from the Earth seems to be a resource ready for exploitation. The interior of the Earth generates 44 terawatts of heat

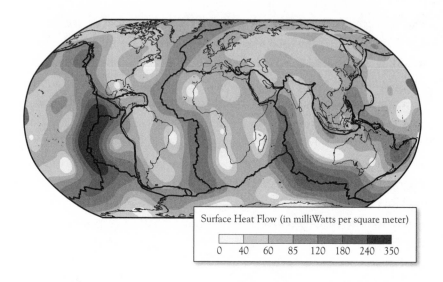

Surface Heat Flow (in milliWatts per square meter)

0	40	60	85	120	180	240	350

Figure III.18. Map of the Earth's heat flow. The average over the surface is about 100 milliwatts (0.1 watt) per square meter.

power, mostly from radioactivity in the Earth's upper crust. That's 44,000 gigawatts. The map in Figure III.18 shows the flow of heat from the inner Earth to the surface.

Geothermal energy has proven very profitable at many places around the world. Iceland generates over one-quarter of its electricity from the heat of the Earth; most of the rest comes from dams. If you include heating, in Iceland geothermal accounts for more than half of all its energy use. A 30-square-mile geothermal area in California called "The Geysers" generates low-cost electricity from the steam that (used to) rise to the surface from that region. Total electric production from geothermal in California is 2.5 gigawatts—about 6% of the state's total electric power.

Forty-four terawatts may seem like a lot of power, but the Earth is big, and except in a few places (such as Iceland), the heat flow is very diffuse. Look at Figure III.18. The average over the surface the Earth is about 0.1 watt per square meter. Compare that to sunlight, which at its peak delivers 1,000 watts per square meter and, when averaged over night and day and north to south, about 250. So solar is 2,500 times stronger than geothermal! If you want to utilize diffuse geothermal energy competitively, you'll need to build a collection system that's 2,500 times cheaper than a solar collector per square meter. Does that sound impossible? I think it is. Geothermal will continue to be valuable in regions where the Earth, on its own, concentrates the heat by factors of 1,000 or more, but not as a worldwide energy solution.

Beware of proposals to take advantage of "geothermal anywhere." Some of them talk about fracking the rock (in the same way that has worked for shale gas)—pumping water down, letting it heat, and then letting it flow back out. The problem with this approach is that heat flows very slowly through rock. Yes, it will reach the water, but once you extract this local heat, it takes eons for the Earth to replace it. Again, look at Figure III.18. To get a gigawatt from a density of 0.1 watt per square meter, you have to collect the heat flow from 10 billion square meters—10,000 square kilometers, nearly 4,000 square miles—a square 63 miles on each side.

The problem is made worse by the fact that low-grade geothermal heat is very difficult to convert to electricity. The temperature of rock increases at a rate of about 30°C per kilometer of depth. The efficiency for extracting energy from heat is limited by the Carnot efficiency equation (which I used previously for solar energy; see Chapter 8, note 4). For water 30°C hotter than the surface, we put in the absolute temperatures (330 for the hot water, 300 for the Earth's surface) to get the maximum efficiency of useful power extraction:

$$\text{Efficiency} = 1 - \frac{300}{330} = 9\%$$

That's pretty low. Instead of 4,000 square miles, which already sounded big, you'll need 44,000 square miles to get 1 gigawatt. Or you could boost the efficiency up to 18% by going twice as deep—to 2 kilometers and water heated 60°C above ambient. Fracking works for natural gas because natural gas is a valuable commodity. Low-temperature heat is not worth nearly as much.

As a result of this physics, geothermal plants to extract energy from bedrock have been failing. One large project by AltaRock Energy was a major initiative by the administration of President Obama. The company received $6 million in government financing, and $30 million from venture capitalists, but in 2009 the project was shut down. A similar project near Basel was terminated by the Swiss because (the government claimed) pumping water underground had triggered local earthquakes. I think it likely that the government used the earthquakes as a face-saving excuse for canceling a project that they had realized was never going to become economically viable. I also expect that fracking for heat will also be opposed by the same people who oppose fracking for natural gas.

Why does optimism about geothermal persist? In 2007, an MIT-led panel concluded that "heat mining" would prove to be a key energy source in the future. Note the use of the term *mining*; the panel understood that geothermal energy is not renewable; extract some heat and then move on to a different location, since it will be hundreds or thousands of years for the rock to rewarm. The MIT report is full of both optimism bias ("the technology will continue to improve") and skepticism bias (they called natural gas "increasingly expensive"—yet soon afterward the price dropped 75%). They called the environmental impacts of heat mining "markedly lower than conventional . . . nuclear power"—a statement with optimism and skepticism biases in the same sentence.

I suspect that part of the continuing enthusiasm comes from the fact that hot rock is found everywhere around the Earth, even under your backyard. Maybe your state has always been jealous of Texas and Alaska and their enormous oil reserves. You missed out on the gold rush and the natural-gas rush, but maybe you can participate in the heat rush!

I expect that geothermal, like tidal power (to be described next), will continue to be economically feasible only in those limited regions where energy is focused by the Earth.

Tidal Power

The power density in tides is low because tides are slow. Mid-ocean tides rise and fall about a meter every 12 hours. That turns out to be an average power density of about 0.1 watt per square meter, comparable to average geothermal power density, and 2,500 times weaker than that of solar. Nevertheless, some tidal power plants have enjoyed considerable commercial success, and as a result, new plants are frequently proposed. As a future president, you must understand why tidal power works sometimes but is not going to be a large contributor to our future power needs.

The most successful tidal power plant so far was built in France in 1966, across the entrance of the Rance River tidal basin. What made the plant commercially successful were the huge tides in this basin, which rise and fall typically 26 feet twice each day, and the large area of water that the power plant could utilize. Why are these tides so large? It is largely a "resonance" effect: left on its own, the water-sloshing frequency in and out of the basin is twice per day. That matches the tidal frequency, so, like a parent pushing periodically on a child's swing, the tides pushing periodically on the basin build up a very high slosh.

The French exploited this high tide by building a half-mile *barrage*, basically a dam, across the river's outlet to the ocean (Figure III.19). The barrage allows water to flow into the 22-square-kilometer basin at high tide, and then it extracts energy from the water as it leaves the basin at low tide—just like a dam that refills the reservoir twice each day. The power plant was considered very expensive in its day, but after 46 years of operation it has paid off all its loans, and the incremental cost is tiny. It currently delivers electric energy at a cost of about 1.8¢ per kilowatt-hour, even lower than nuclear (2.5¢ per kilowatt-hour in France). It delivers a peak power of 240 megawatts, but the average is much lower, about 100 megawatts.[45]

Why don't we build more such plants? There are two basic reasons. The first is that such high tides are very rare around the Earth.

Figure III.19. The dam at the inlet of the Rance River in France. The flow of tides at this facility generates an average electric power of about 100 megawatts.

The power available depends on the square of the tide height.[46] For New York City or San Francisco, a typical tide height is 6 feet (rather than the 26 at Rance), so the power available for the same area is down by a factor of $(6/26)^2 = \frac{1}{19}$. That would make the difference between a 100-megawatt power plant and a 5-megawatt plant. The second reason we don't make wider use of tidal power is that a barrage can do substantial environmental harm, not only by changing the access of fish in and out of the basin, but also by changing the salinity of the basin. Back in 1966 France was a relatively poor country, still not fully recovered from the devastation of World War II, and such environmental considerations were not topmost on its list of worries.

A less harmful way to build a tidal plant is to use submerged power generators. This is what was done at the entrance to Kaipara Harbour in New Zealand. Again, it makes economic sense only if the tides are large, and at this entrance they are 7 feet high. Power is generated by 200 submerged generators producing 1 megawatt each.

You can think of them as underwater turbines, in which flowing water spins a rotor that makes electricity. The New Zealanders are proud that their generators are invisible and silent from the surface. The estimated cost of these generators is $600 million for the first 10 years. That's $3 per installed watt—a reasonable value, but not cheap. Maintenance costs are still not known, and I suspect they will be very high. The submerged-generator approach does less environmental harm than does a barrage, although some fish are probably killed.

The biggest tidal station in the world (just barely) is the Sihwa Lake Tidal Power Station in South Korea, built at a location that has 18-foot tides. The plant was commissioned in 2011, and it has a 254-megawatt peak capacity (designed to be just above the 240-megawatt peak output of the Rance plant). It, too, uses a tidal barrage, but unlike the one in France, the Koreans claim it will be environmentally beneficial. A seawall previously built at the same location in 1994 for flood control and agricultural purposes had caused a buildup of pollution. The barrage will restore the flow and, the South Koreans hope, restore the environment. It's not meant to be better than nature, but it should be an improvement over prior human damage.

And, not surprisingly, a power plant has been built at Canada's famous Bay of Fundy to take advantage of its monstrous 56-foot tides (also caused by a resonance effect). But the Bay of Fundy has a very wide mouth, so the power plant intercepts only a small fraction of the tidal power and generates only 20 megawatts.

Tidal plants have been proposed for the United States. Recently, the Golden Gate Energy Company signed a statement of collaboration with the city and county of San Francisco, and with Pacific Gas and Electric, to consider putting generators underwater near the entrance to the Golden Gate, where water flows in and out of the huge San Francisco Bay twice each day.

In the end, tidal power will be limited to local regions that can get local benefits. Tidal power plants require places that have uncommonly large tides or that otherwise (like the Golden Gate) focus the energy of tides through a narrow channel. They will not be important for the power needs of the overall United States or the world. The Rance Tidal Power Station, second largest in the world, supplies only 0.01% of the French power demand.

Wave Power

Just as winds are driven by solar heating, waves are driven by wind, and the power in them seems enormous. Global wave power hitting the coasts adds up to 3 terawatts; that's 3,000 gigawatts. Total world energy use is currently only 5 times higher than that.

The problem is that waves are just not very high, averaging only about 1 meter around the world. If you were to intercept 100 meters of these waves and extract all the power from them, you would get only 1 megawatt. Recall that just one large wind turbine can deliver 7 megawatts. How can waves in corrosive salt water compete with wind as a source of power? My opinion is that they can't.

Figure III.20. Pelamis prototype machine at the European Marine Energy Centre.

Despite all its problems, wave power has been tried. Portugal built an experimental device called the Pelamis Wave Energy Converter (Figure III.20). It consists of a series of 750-ton floats that flex as waves pass under them. The device is 140 meters long, and it can generate 750 kilowatts, about 5 megawatts per kilometer, meaning that it is successfully converting about half of the power in a typical 1-meter-high wave.

If such power were cheap, then it might be of some interest. But the machine has to operate in corrosive salt water, and anyone who has served in the navy or who owns an oceangoing boat knows what headaches that can cause. The cost of a single Pelamis unit is estimated to be $5.6 million—about $7.50 per installed watt. That wouldn't be too bad if there were no maintenance costs, but the sea environment is tough. Wave power will never be easy, because the sea is severe and the power density low.

16

ELECTRIC AUTOMOBILES

WE ARE addicted to gasoline automobiles for some very good reasons. Consider the following:

- **Fill rate.** When you fill up the fuel tank, you are transferring gasoline at a rate of about 2 gallons per minute. Given the energy density of gasoline, that is equivalent to an amazing 4 megawatts.[47] However, an internal combustion engine is only 20–25% efficient (electric engines are 80%–90% efficient), so that means that *useful* energy is transferred at a rate of only 1 megawatt. Even so, that is a huge number—enough electricity for 1,000 small homes.
- **Range.** After a 10-gallon fill-up that takes only 5 minutes, you can drive 300 miles in an average US auto.
- **Residue.** When you've used all the energy in your tank, there is no ash, no residue, nothing to clean out. (I grew up in the Bronx, where we burned coal for heat. It was as much effort shoveling the ash out of the furnaces as it was shoveling coal in.)
- **Cost.** At $3.50 per gallon, and 35 mpg, the cost of fuel for one mile is only 10¢. This is so cheap that many people choose to move to homes that are far from their workplace and commute large distances every day, suffering traffic jams and long travel times, but getting the benefit

of a better choice of home. According to a survey of car commuters made by the market research firm TNS in 2005, the average auto commute takes 26 minutes each way to travel 16 miles. That's about a gallon of gasoline per day, or about $3.50. That's so cheap that we spend more on luxury and comfort than we do on the fuel. It's a bit different in Europe, where high taxes push up the fuel price to twice that in the United States.

- **Emissions.** The emissions from gasoline are primarily carbon dioxide and water vapor. (Soot and nitrous oxides are now largely under control in the United States.) Those are the same gases that we breathe out. And prior to our worries about global warming, carbon dioxide was considered benign—it helps plants grow.

These features of gasoline automobiles (except for the last one) make it very difficult to change the way we travel in any significant way.

Now here are some additional key facts you need to remember when you think about the future of automobiles. (Some of these I mentioned previously, but I bring them together here.)

- **Global warming.** US automobiles have contributed about $1/40°C$ so far. With feasible mpg standards, and no other changes, they will likely contribute only another $1/40°C$ over the next 50 years.
- **Trade deficit.** Half of the US trade deficit comes from importing oil, which is used mostly for transportation.
- **Battery-auto range.** A kilowatt-hour of electric energy will enable you to drive 2–3 miles. At a kilowatt (typical average house electric energy rate), it takes a battery an hour to charge to that level.
- **Battery cost.** For an all-electric auto, electric energy is not the significant cost. Much more important is the expense of replacing the batteries after 500 charges.

Let's start with the first point: that US automobiles have contributed only $1/40°C$ (0.045°F) to global warming over the past 50 years. I calculated this number in Chapter 13, and I repeat the calculation in note 48 . How much will US automobiles contribute over the next 50 years? The number of vehicle miles driven, according to the estimate of the Energy Information Administration, will probably increase by 60%. That would increase the expected future warming contribution from US autos from $\frac{1}{40}$°C up to $\frac{1}{25}$°C. If the government imposes feasible mpg standards, then the increase could be held to $\frac{1}{40}$°C.

If US autos contribute so little to global warming, why are we working so hard to improve mileage and to develop alternative technologies, such as electric autos and hybrids? The answer, of course, is that we aren't. The number of hybrids sold in the United States in 2010 was 28,592, representing 2.5% of all US vehicle sales. The number of all-electric vehicles so far is much smaller. The number of all-electric cars may grow, but there are salient reasons why they probably will not.

The Electric Auto Fad

Calling the electric auto a fad is likely to draw more ire than many other things I say in this book. *Fad?* Yes. Forgive me for being a bit more pugilistic here than elsewhere, but it seems necessary. In virtually no other area of energy science and policy is there so much unchallenged hype as in the field of electric cars. And in no other field is there a group more enthusiastic, more optimistic, more . . . fanatical . . . than those excited about electric cars.

Note that I am not now talking about hybrid autos, those that use both electric and gasoline in combination. Those have true value and a strong future; I expect that in a decade or two virtually all of our automobiles will be hybrids. It is the all-electric cars like the Tesla Roadster and the Nissan Leaf, and even the Chevy Volt and the plug-in hybrids when operated in all-electric mode, that are part of this fad. I expect the interest in such cars to be short-lived.

Three fundamental problems must be solved if all-electric cars are to become widely used in the United States: energy density, cost, and recharge time. The energy density problem comes from the fact that electric batteries can store only about 1% of the energy per pound that gasoline can. Compensating a bit for this shortcoming is the fact that electric engines can operate about 4 times more efficiently than the internal combustion engine used in gasoline autos. Therefore, the useful energy density for batteries is closer to 4% that of gasoline. That means you need a lot of batteries, unless your range is very low.

Now let's look at price. We can assume the electricity cost is low; the main cost for electric autos is in the batteries. Beware. Many articles about electric cars look only at the electricity cost and ignore battery replacement, so they vastly underestimate the true cost of operation.

The batteries used in these cars are lithium-ion, which range widely in price—from $30 to $150 per pound. The replacement battery for my laptop is $120 for one pound. Generally speaking, the cheap batteries are not a bargain, because of their typically shorter lifetime. I no longer buy them for my digital camera because I've learned that the number of recharges they deliver is often less than 100. The good batteries (such as the one for my laptop) are guaranteed to last 400 recharges. Some battery manufacturers claim that their batteries can recharge 1,000 times or more, but be wary. For my calculations, I'll assume that the batteries bought in bulk cost $40 per pound and can be recharged 500 times. It takes 25 pounds of batteries, worth $1,000, to hold one kilowatt-hour of energy; that's $1 per watt-hour. Since they need to be replaced after 500 recharges, the batteries cost $1 for 500 watt-hours, equal to $2 per kilowatt-hour delivered. Compare that to 10¢ per kilowatt-hour from your wall plug. Replacement cost overwhelms electricity cost.

TESLA ROADSTER

The first electric-auto example we'll look at is the Tesla Roadster. It has 1,100 pounds of batteries, costing (at $40 per pound) $44,000. The batteries comprise 44% of the weight of the car. The advertised range is 250 miles, but that would require driving the car at an average of 16 horsepower. Owners report that a more typical range is

125 miles, and that number agrees more closely with the range per kilowatt-hour reported for the Chevy Volt. With 500 recharges, you could go 62,500 miles before needing to replace the batteries. Tesla guarantees the batteries for only 36,000 miles or 3 years, whichever comes first. The major cost per mile to operate the car is not in the cost of the electricity, but in the prorated cost for battery replacement. That comes to $44,000 for batteries ÷ 62,500 miles = 70¢ per mile.

Compare that to a conventional auto with an internal combustion engine that gets 35 mpg, with a gasoline cost of $3.50 per gallon, and a cost per mile of $3.50/35 = 10¢. The Tesla Roadster is 7 times more expensive to operate per mile. Of course, many people who buy the $110,000 Tesla Roadster don't do it to save fuel costs.

You might argue that you never will replace the batteries; you'll just sell the car. But the Roadster won't be worth much after 62,500 miles if the prospective buyer will have to install $44,000 worth of new batteries. Tesla says it will replace the batteries for $36,000. If so, it may be counting on a drop in battery prices (more on that in a moment) or on making no profit (or taking a loss) on replacements.

Tesla stopped producing the Roadster in late 2011. Tesla's reason: not a big enough market. Another possible reason is that the batteries were a lot more expensive than $40 per pound (my computer battery costs $120 per pound). Add in the fancy body, and you'll conclude that Tesla was losing substantial money on each car.

Why would Tesla do that? Maybe the company was optimistic that the price of batteries would come down, and thwarted when it didn't. Or maybe it was just trying to establish the Tesla name for sales of future smaller and less expensive cars.

CHEVY VOLT

The Chevy Volt has 375 pounds of batteries that cost $15,000 (at $40 per pound). Yet GM claims that a single charge will take you only 40 miles. Note that such a short range is completely in line with what the drivers of the Volt actually report. Five hundred charges will take you 20,000 miles. I suspect that GM is being candid about this number because the Volt has a gasoline backup that guarantees you will not get stuck if the battery runs down. The cost of batteries is $15,000/20,000 = 75¢ per mile. A General Motors manager has stated publicly that the company is making no profit on selling these automobiles at $40,000 each. That seems plausible, given the huge battery cost.

NISSAN LEAF

Finally, consider the Nissan Leaf: 400 pounds of batteries cost about $16,000. Each charge is advertised to take you 100 miles, but according to the US Environmental Protection Agency the true range is 73 miles. (Both numbers seem oddly higher than the 50-mile advertised range for the Volt, which has nearly the same weight in batteries.) For 500 charges, that adds up to 36,500 miles before you have to buy a new set of batteries for $16,000. The cost per mile is 44¢.

In late 2011, the Nissan Leaf cost $34,700. A home slow charger is an additional $2,000 (a fast charger costs $40,000—too expensive for most people). The comparable Nissan Versa Compact, an ordinary gasoline car, costs about $14,000. The difference is $20,700. The Versa gets about 35 miles per gallon; at $3.50 per gallon, it costs you 10¢ per mile to drive it, versus 44¢ for the Leaf. Don't buy an all-electric car hoping to save money.

The $7,500 subsidy offered by the US government reduces the purchase price difference, so the Leaf is *only* $13,200 more expensive than the Versa. But the subsidy doesn't change the cost per mile. Drive 109,500 miles (requiring three battery replacements), and at a 34¢ added cost per mile the Leaf driver will have spent $37,230 more in operational and fuel costs. Add that to the purchase price difference of $13,200, and the Leaf owner—for similar performance—paid $50,430 more than the Versa owner.

I expect the interest in all-electric automobiles to die soon, when the batteries need replacement and the owners discover how much they cost.

Plug-in Hybrid Electric Vehicles

The most famous of the hybrid automobiles, the Toyota Prius, can travel only 4 to 6 miles on its small battery. The most recent Priuses have an EV ("electric vehicle") switch that allows this mode—useful in emergencies or a trip to the corner grocery store. If you want more electric range, you can buy a kit to convert the car into a plug-in hybrid electric vehicle (PHEV) for $10,875. In that kit you get a battery that holds 5 kilowatt-hours of electricity. That means you're paying about $40 per pound of battery. A *Consumer Reports* investigation found that over 3 years, the kit would allow enough all-electric operation to save a typical driver $2,000 in gasoline. But then the battery would have to be replaced for an additional $10,875. If you own a recent Prius and are interested in saving money, never push the EV switch.

Toyota is selling plug-in Priuses in Japan. The price of gasoline (when I last checked) was 150 yen per liter—about $2, which translates to $7.60 per gallon. That means that in Japan, the extra cost of running a plug-in hybrid is not as much as in the United States. After 3 years you would save $4,000 for your $10,875 expenditure.

Proponents of electric autos might argue that my claimed cost of batteries is too high. I disagree. Think about the rough treatment these batteries are subjected to in an auto—temperature change, rain, vibration, bouncing. Only very high-quality batteries subjected to rigorous quality control measures can last 500 cycles under such conditions. And I don't expect the price of batteries to come down, at least not rapidly. The battery kit for the Prius cost $10,000 in 2008, and $10,875 in 2011. That's only 3 years, but the kit uses a very popular and highly hyped battery called the A123, and its price has clearly not dropped in that period.

Lead-Acid Batteries

Remarkably, the best bet for the electric car of the future may be the old-fashioned lead-acid battery we use to start our ordinary cars. Compared to lithium-ion batteries, lead-acid batteries hold only half the energy per pound, but they've been around for a long time, and they're cheap. You can get a good 50-pound battery for $100; that's $2 per pound, compared to $40 per pound for lithium-ion. Assuming 500 recharges, the replacement cost is 5¢–10¢ per mile. Add the cost of electricity (5¢ per mile), and you get a total similar to the price of gasoline.

The catch is the limited range. The 2006 movie *Who Killed the Electric Car?* was about the GM EV1 battery auto that used lead-acid batteries to give a 60-mile range. Eventually the EV1 battery was replaced with a nickel–metal hydride (NiMH) battery, which extended the range to over 100 miles. Unfortunately, NiMH's are about 10 times as expensive as lead-acid batteries for the same energy storage, bringing the price of the batteries up from $2,000 to nearly $20,000 and increasing the cost per mile by a factor of 10, making the car completely uneconomical. And still the batteries gave only a 100-mile range.

Some people say that General Motors killed the EV1 because of pressure from oil companies that feared it would put them out of business. But the numbers suggest that GM correctly determined that the lead-acid car had too limited a range, and that the NiMH car was too expensive to operate, so the market would consist solely of electric-auto enthusiasts who could afford the extra cost; that market would remain too small to make the car viable. Who killed the electric car? It wasn't a who; it was a what: the cost of batteries.

Recharge Time

There is a second problem with electric autos: the time it takes to fill 'er up. For example, consider the Tesla Roadster. It was designed to recharge in 3.5 hours with a "High Power Wall Connector." This device takes 240 volts and 70 amps; to calculate the watts, multiply those two numbers together: 16.8 kilowatts. That's about 10–16 times higher power than the average electric power used by a typical US home. The Roadster's 1,100-pound battery holds about 66 kilowatt-hours of energy. To charge the battery at 16.8 kilowatts would take 66/16.8 = 3.9 hours. Could you do it faster at a service station? Yes, but not much faster. As you know if you've ever charged an ordinary lead-acid battery, or a computer battery, rapid recharging can reduce the battery lifetime. The new Tesla Model S boasts that at a high-voltage service station, the auto could be recharged in under an hour.

Some people believe that the solution to the long charging time is battery exchange. Rather than charge the battery, replace it at a trade station for one that is already charged. Making such an exchange with 500 pounds of batteries is possible, but not easy. Recall that the maximum range is 40 miles for the Volt and 73 miles for the Leaf, so you would have to stop at a station every hour or so. Because your batteries are used, when replaced you would have to pay the depreciation, amounting to 44¢–75¢ per mile driven. One benefit of an exchange system in my mind is that drivers would be made intimately aware of the high cost of driving all-electric vehicles.

Effective Miles per Gallon

Many of the new electric cars boast incredible *miles per gallon equivalent* (mpg-e). According to the EPA, the Nissan Leaf gets 99 mpg-e, and the Chevy Volt gets 93 mpg-e. Those high numbers are proudly displayed in the showrooms. The great Toyota Prius gets only 50.

The EPA also gives the cost of fuel for 15,000 miles: $561, $594, and $1,137 for the Leaf, Volt, and Prius, respectively.

Those numbers sound fantastic. For the Volt, it means a fuel cost of 4¢ per mile—compared to 10¢ for a 35-mpg conventional gasoline car. Alas, these numbers are enormously misleading, and they encourage people to buy cars to save money when in fact, if the battery replacement costs are included, those cars will wind up costing them much more. For the Volt, the truly honest number would be 4¢ per mile for the fuel + 75¢ per mile for battery replacement.

One gallon of gasoline can be burned to produce 33.7 kilowatt-hours of energy, so the EPA defines that as the conversion factor. The reason that the mpg equivalent is so misleading is that most of the energy losses take place before the electricity "fuel" gets into the automobile. Once there, batteries are efficient, and so are electric motors. The losses—in the generator and the transmission lines—are all external and not included. That omission gives the misimpression that electric automobiles are far more fuel-efficient than are ordinary gasoline autos.

Let's do a calculation. Here are the efficiencies:

- Fuel burned in the best modern power plants: 45% efficient
- Electric transmission line efficiency: 93%
- Batteries charged and discharged: 80%
- Electric motors: 80%.

Multiply these together to get 27% efficiency for electric vehicles—slightly higher than the 20% obtainable by an internal combustion engine, but not wildly more efficient, as the mpg equivalent value seems to suggest. And if the electricity was generated by coal, the CO_2 produced is much higher for the electric "zero emission" auto.

Ordinary Hybrids

What about the ordinary hybrid, such as the Prius? I will argue that it should *not* be classified with the other electric automobiles—that unlike them, the ordinary hybrid does indeed make economic sense. According to the EPA, the Prius gets 50 miles per gallon of gasoline. That is, in fact, what I get if I drive my Prius with a little bit of care, keeping my speed at about 65 mph even when I can go faster, and only occasionally passing other cars.

The Prius uses a very small NiMH battery with an energy storage capacity of about 2 kilowatt-hours,[49] enough for 4 to 6 miles of range. The battery is used when the car is initially accelerating, and it recharges when the car slows down; instead of dumping the kinetic energy of the car into heat, as brakes do, the Prius uses "regenerative braking": a wheel turning an electric generator requires force, and that force (rather than friction) can slow the car. If the battery runs down, then it is recharged by a generator run by the gasoline motor.

Most important, the Prius doesn't use its battery at all when the car is coasting efficiently along at 65 mph. It is a very intelligent design: use the expensive-to-replace battery only during those short moments of the driving cycle when an ordinary car is most inefficient. If you're taking a long trip, you might cover 350 miles but age the battery the equivalent of only 10 miles. It means that the main expense of battery-driven automobiles, battery replacement, is minimized, yet it saves gasoline when the car would otherwise be wasting it. You undo this smart engineering if you convert the car to a plug-in.

Consumer Reports tested the battery on a Prius that had been driven 200,000 miles in 8 years and concluded that the battery had shown little degradation, if any. However, that conclusion can be misleading, since a car that was driven 25,000 miles per year probably spent most of its miles on the highway where the battery was hardly used. The Prius replacement battery from Toyota costs about $2,200. For the 2-kilowatt-hour battery, that's $1,100 per kilowatt-hour, comparable to what I assumed[50] for the all-electric Roadster and Leaf.

Plug-in vehicles currently earn a $7,500 tax credit. President Obama's 2012 budget request transforms this credit to a point-of-sale rebate, so that the refund is issued immediately. This rebate could offset the $8,000 additional cost that the consumer would otherwise have to bear after 3 years of driving (based on the *Consumer Reports* estimate). The rebate makes some sense for energy security, since it lowers the need for imported oil, but it is a very expensive way to accomplish that. The law's true purpose, to encourage plug-in hybrids, is poorly considered. It would be much better to encourage ordinary hybrids, or to simply raise the mpg standards (through limits on the Corporate Average Fuel Economy, abbreviated CAFE) required of the automobile industry.[51]

The bottom line is that the Prius ordinary-hybrid design makes a great deal of sense, and I expect that such hybrids will prove in the long term to be a much more popular option than the all-electric auto. If you own a Prius and care about cost, do not convert it to a plug-in hybrid.

Battery Challenges

If batteries get a lot better, then all-electric vehicles can have a future. Here's a summary of the challenges that such vehicles must address to become truly competitive:

- Energy per pound is 25 times worse than that of gasoline.
- Cost per mile is 5–8 times higher than that of gasoline (including cost of electricity and battery replacement).
- Storage tanks take 10 times more space (for the same range).
- Refill time is not minutes, but an hour at the service station and much longer at home.
- Initial capital cost of the batteries is measured in tens of thousands of dollars.

Batteries *are* getting better, but slowly. The difficulty is not in battery chemistry; we understand that very well. It is in the nanotechnology of the electrodes and electrolytes. I think it unlikely that improvements in the next 20 years will be sufficient to make batteries economically competitive with gasoline, particularly if gasoline mpg improves.

All-electric cars using lead-acid batteries are already competitive, if Americans are willing to accept a 40-mile range. Why do we insist on long ranges for our cars? Part of the reason is undoubtedly habit and our love of long trips. Part of it is the fear of running out of gas and being stuck far from a filling station—"range anxiety" already being reported by Leaf owners. Part of it is our simple dislike of having to stop to buy gas. But much of the world does not yet share this US addiction to long range. In China, India, and Africa, where many people haven't previously owned a car, the idea of a 40- to 60-mile range can appear attractive. And in many of these countries the price of gas is higher than in the United States. For this reason, I suspect that the true future of electric autos will be in the developing world using the cheapest of all batteries: lead-acid.

What impact do battery-operated cars have on global warming? Electric autos currently derive their energy primarily from fossil fuels. An auto running on electricity from a coal-fired power plant, when you include power line and other losses, dumps more carbon dioxide into the atmosphere than does a gasoline auto.

Does the efficiency gain justify the high cost? Or would money spent on expanding electric vehicles (for example, subsidies or rebates) be better spent on improved infrastructure (such as power lines), encouragement of solar, wind, and nuclear (for instance, through legislation that creates a market for such fuels, such as California AB32), and on improvements in the mpg (CAFE) standards? That is a judgment you will have to contend with when you are president.

17

NATURAL-GAS AUTOMOBILES

THANKS to the US natural-gas windfall, it makes sense to look even more seriously at the possibility of converting our automobiles to run on natural gas. The best candidate is compressed natural gas (CNG) rather than liquefied natural gas, simply because the liquid must be kept below –259°F to keep from boiling. Gasoline costs the consumer 2.5 times more for the same energy, and the ratio is rising.[52]

Natural gas can be readily compressed to 250 atmospheres—that is, to a density 250 times greater than normal—and this high-pressure gas can be carried in steel or fiber composite tanks. When compressed to this level, natural gas has an energy of 11 kilowatt-hours per gallon, only a third as good as gasoline, but 10 times better than lithium-ion batteries, and 4 times better than hydrogen compressed to the same pressure. It's a natural.

Since 1998, Honda has been selling a natural-gas vehicle in the United States, the Civic GX (Figure III.21). It costs $25,490 (manufacturer's suggested retail price). The ordinary Civic sedan, with comparable performance, costs $15,805. The $9,685 difference is a major disincentive. Why is the GX price so high? The biggest contributor is a device that Honda calls "Phill" that it includes with every car. Phill is a compressor that can take your home natural gas

Figure III.21. The Honda Civic GX, a natural-gas auto.

and compress it into the GX tank in a few hours. Let Phill fill your tank. Sold separately, it costs $6,200, including installation in your home. The high-pressure tank for the car adds another $1,500. The carbon fiber composite pressure tank typically weighs 5 times more than the weight of the gas it holds.

Clearly, a natural-gas auto is much cheaper if you don't have to buy the compressor or if you can share the compressor with other people. For this reason, most of the natural-gas vehicles have been parts of fleets, large numbers of vehicles that can use the same refill facility, or have been used in regions of the world that have a natural-gas fill station infrastructure. You can look for such a station near your home by checking the website of the Alternative Fuels and Advanced Vehicles Data Center:

www.afdc.energy.gov/afdc/fuels/natural_gas_locations.html

where you'll find a map for your state. California has over 200 natural-gas stations; New York, over 100. Figure III.22 shows the locations of stations in the New York City area in 2011. In areas with lots of stations you may not need the home compressor, and the economics of natural-gas automobiles then becomes more attractive.

You can convert an existing auto to natural gas by using a conversion kit that costs between $2,000 and $4,000, plus installation.

The kit adds a new tank (typically taking up trunk space), the old gasoline tank remains, and you can run the car in a dual mode: natural gas when it is available, gasoline when there are no nearby natural-gas stations.

The cost of the natural gas to drive one mile is about 4¢. Compare that to the current cost of 10¢ per mile for gasoline (more if gasoline prices rise) or to the battery replacement cost of 44¢–75¢ per mile. Natural-gas tanks are much cheaper than batteries, and unlike batteries they do not require replacement after 500 refills.

Here's a simple cost estimate. Suppose you have natural-gas pumping stations nearby, so you don't need Phill; you install a conversion kit and tank for $3,000; and you drive 15,000 miles each year. You will save 6¢ per mile (the difference between 10¢ and 4¢), and every

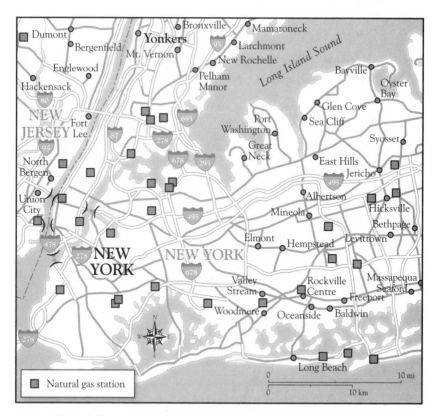

Figure III.22. Natural-gas stations in and around New York City.

year that will add up to $900. That's a return on your investment of 900/3000 = 30%. You might want to subtract 10% for depreciation, but that still is a 20% return on your investment.

Include Phill and the investment doesn't look very good. You still save $900 per year, but your cost was $3,000 for the conversion + $6,200 for Phill = $9,200, and your yearly return is 900/9200 – 10% (depreciation) = –0.2%. Not so good—but not bad either; you break even, you can refill your tank at home, and you have a car you can show off.

The economies of natural gas are stronger in countries that have higher gasoline prices—assuming the natural-gas prices are not higher too. These countries use mainly shared filling stations. As a result of these economies, in the world there are over 12 million natural-gas vehicles, although only about 150,000 in the United States (mostly taxis, buses, and municipal vehicles). Pakistan has nearly 3 million natural-gas vehicles; India has over 1 million. The developed world (members of the Organization for Economic Co-operation and Development, or OECD) has about a half-million.

Natural gas has other advantages. Because the tank is so sturdy (to hold the pressure), a little incident like a head-on car crash hardly dents the tank; according to its proponents, natural gas is safer than gasoline. It is also safer because it is lighter than air, so if it does escape, it quickly rises (unlike pools of gasoline). It also requires a higher ignition temperature than gasoline, so even if confined it is less likely to ignite. Finally, the use of natural gas seems to reduce wear on the engine. This effect could help slow the depreciation rate of the vehicle. There are all sorts of reasons why natural-gas automobiles make sense. Even synfuel has difficulty competing with natural gas.

The main downside of natural gas is range. At 250 atmospheres (3,600 pounds per square inch, or psi), for the same-volume tank natural gas will take you only a third as far[53] as gasoline does. In the Honda Civic GX, the natural-gas tanks are hidden under the floor, so you still have reasonable trunk space, and they have enough volume to give a nominal 250-mile range. But the car is heavier because of these tanks, and that added weight reduces its miles per gallon.

In 2003, President George Bush initiated a program to build hydrogen fueling stations around the United States. Hydrogen currently sells for about $8 per kilogram; as a fuel it costs 27¢ per kilowatt-hour. In contrast, natural gas costs 4¢ per kilowatt-hour delivered. And it can store equal energy in a quarter of the space that hydrogen requires. Instead of hydrogen, President Bush should have touted natural gas.

18

FUEL CELLS

FUEL cells have been proclaimed as the eventual replacement for batteries for over a hundred years. Now they're being proclaimed as the eventual replacement for standard power plants and for automobile engines. Is that optimism warranted?

If you've ever seen a demonstration of the electrolysis of water, you've seen a fuel cell operating backward. In electrolysis, electric current is passed through water, releasing hydrogen gas at one end and oxygen at the other, as illustrated in Figure III.23. To turn an electrolysis cell into a fuel cell, replace the battery with a wire and supply hydrogen and oxygen gases at the appropriate ends. Out will come electricity. The way it works, chemically, is that some hydrogen atoms lose an electron, drift through the water, and react at the other electrode with oxygen to form water. The two electrodes are left charged, and you can get the electricity out by connecting them with a wire. Then use that electricity to run a lamp or a motor. The key fact is that the fuel cell is unchanged and doesn't need to be recharged. Except for a few engineering details, you just supply the fuel (hydrogen and oxygen), remove the product (excess water), and out comes electricity.

You could instead just burn the hydrogen and use the resulting

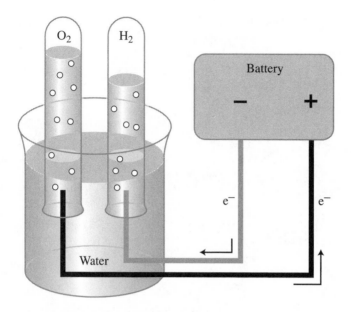

Figure III.23. Electrolysis. To change this device into a fuel cell, simply replace the battery with a lamp.

heat to generate electricity, but the efficiency of that approach is low, typically 20%–35%. Part of the excitement over fuel cells comes from the high theoretical efficiency of 83%, although this value is rarely reached in practice. Fuel cells also have the advantage of being relatively simple, small, clean, and quiet.

Fuel cells can also work with methane and air. If operated at high temperatures, the electrolyte can be solid, typically a ceramic. The Bloom Energy fuel cells pictured in Figure III.24 are examples installed at Fireman's Fund headquarters in California. I find their stark simplicity beautiful, reminiscent of the monolith from the movie *2001*. The cells can be placed close to where the power is needed, right on the grounds of the business or factory, thus avoiding the typical 7% electricity transmission loss. If they're nearby, you can use the heat generated (the portion of the energy that is not converted to electricity) to warm your buildings; this "combined heat and power" means that virtually none of the fuel is wasted.

The fact that the waste heat can be used has led to some mislead-

Figure III.24. Bloom Energy methane fuel cells installed at Fireman's Fund headquarters in California.

ing claims of superhigh efficiency, 80%–90%. Those claims have wowed some reporters, who thought it was enormously better than the typical 40% efficiency achieved in natural-gas turbines. But heat energy is not nearly as valuable as electric energy, so it is misleading to lump the two together. By such terminology, all engines are 100% efficient if you use the waste heat to warm a building. Moreover, you can claim 100% efficiency in burning natural gas for heat. If you use natural gas to drive a heat pump, you can claim efficiencies greater than 100%.[54] Because of this confusion, many people thought that the Bloom Energy fuel cells represented a fundamental breakthrough in technology, when they were in fact just an incremental improvement (because of their proprietary ceramics) over the prior art.

To be practical, the cell not only must be efficient but must deliver high power, keeping the cost reasonable, making sure that the electrodes don't corrode or foul, and that the material between the electrodes, the electrolyte, remains unchanged. The structure used also must resist the cycles of heating and cooling that stress fuel cells. Cells that operate at low temperatures typically have to be loaded with a catalyst, something that makes the chemical reactions take place faster. The best catalyst, unfortunately, is very expensive: platinum, the same material that is used in automobile catalytic converters. The current (2012) price of platinum is $1,600 per troy ounce, or $51 per gram. For even a moderate fuel cell, one that could be used to run a laptop computer, the cost of the platinum is several hundred dollars. For certain applications, such as in space vehicles or to support a battlefield soldier, that high cost isn't an impediment, but to compete with other ground energy sources, it is a severe limitation.

Because of its expense, there is an enormous ongoing research effort to find a substitute for platinum. Catalysts typically work by adsorbing onto their surfaces the molecules that need to react, and doing so in a way that places the molecules close to each other and in an optimum orientation to enhance the chemical reaction. This may sound to you like nanotechnology, and indeed that is the field in which much of the work is done. The key, however, is not just to make a better catalyst, but to make one that is cheap enough to be worth using.

To avoid the expense of platinum, an alternative is to go to high-temperature operation, typically about 1,000°C (1,832°F). At high temperatures two things happen: First, ordinary fuel cell chemistry speeds up. Second, a new chemistry happens that is enormously useful: methane reacts with water to release hydrogen gas and carbon monoxide. That means you can introduce methane into your cell rather than the far more expensive hydrogen. In effect, you're separating the hydrogen within the fuel cell itself—a process called "reforming." That is why the Bloom Energy fuel cells can operate with natural gas.

The most successful of the high-temperature fuel cells uses an exotic ceramic for its electrolyte: YSZ.[55] This is a miraculous material that, when heated, allows oxygen ions (each carrying two electrons) to flow through it, while not allowing free electrons to pass. That's possible because (to the surprise of many non-physicists) heavy ions are actually "smaller" than electrons in a quantum mechanical sense, so they can sneak through the crystal structure. That's just what is needed for a fuel cell. The oxygen ions drift through the YSZ and chemically react with the hydrogen on the other side, creating water and leaving their excess electrons (which stick to oxygen but not water) on the metal electrode. Since the electrons can't move through YSZ, they return on the external wire, providing power.

Alas, platinum is too expensive for an automobile fuel cell, and YSZ cells too hot, so materials scientists are searching hard for a replacement material. One possibility is cerium gadolinium oxide, or CGO. Do all these materials—cerium, gadolinium, yttrium (the Y in YSZ)—seem strange to you? It is a remarkable aspect of modern materials science that scientists seem to test *everything*, and often discover truly remarkable and marketable properties. The material used in the Bloom Box is still secret, but experts have speculated that it contains YSZ with cerium added.

Fuel cells have enormous potential benefits. Why are they not more widely used? The obvious answer is cost. Although Bloom Energy claims that its fuel cells can produce energy for 14¢ per kilowatt-hour, it has not released the data to substantiate that claim, and it may be including the heat produced. Many speculative companies make extraordinary claims based not on their current costs but on projections of what those costs could reduce to in the future if their sales were to grow to a billion dollars annually. In an April 2010 interview with *Newsweek* magazine, K. R. Sridhar, the CEO of Bloom Energy, admitted that the current cost of the Bloom fuel cell was $7–$8 per installed watt.[56] That makes it over 7 times more expensive than a natural-gas turbine plant. These numbers are consistent with the numbers in a November 2010 report issued by the Energy Information Administration: "Updated Capital Cost Estimates for Electricity Generation Plants."[57] According to the EIA report, fuel cells cost $6.83 per installed watt, whereas a more conventional (but combined cycle) natural-gas turbine generator plant costs only 99¢ per watt.

For commercial power production, the major manufacturer of fuel cells is United Technologies Corporation, the company that originally supplied fuel cells for our space missions. UTC has installed over 75 megawatts of phosphoric acid fuel cells across 19 countries. Note how small that number is; the total power of these fuel cells is 7.5% of a single typical nuclear or coal power plant. These plants require hydrogen, which is obtained by a separate reformer that reacts natural gas with water. The UTC fuel cells achieve an electric efficiency of about 40%.

How can fuel cells compete commercially when they are so expensive? Bloom Energy CEO Sridhar admitted in the same *Newsweek* interview, "Right now we are only economical with subsidies." The Energy Policy Act of 2005 grants a tax credit of 30% or $3,000 per kilowatt and accelerated depreciation. California has paid over $630 million in incentives for fuel cells, resulting in over 350 megawatts of generating capacity. Currently, 28 states have laws called "renewable portfolio standards" (RPSs) mandating that their utilities provide a certain fraction of their power from renewable energy sources; many states include fuel cells in this list because of the fuel cell capability to use biomethane. (I argued in Chapter 13 that not everything "bio" should be considered renewable, at least from the carbon dioxide emission point of view, since the alternative to consumption is to leave the carbon buried—that is, sequestered.) An RPS is effectively a subsidy, since utilities are guaranteed a profit, and when they use expensive fuel cells they can raise the rates on their customers. Politicians like RPSs because they provide funds without appearing to be government taxes.

Because of the extreme difficulties in making a mobile methane fuel cell—one that can endure the vicissitudes of weather, bumps, and acceleration—it is very unlikely that fuel cells will be practical for use in automobiles in the next decade or two. The Honda Clarity (Figure III.17) runs on hydrogen, not methane; it can run at relatively low temperatures but it requires expensive platinum. Toyota plans to produce a hydrogen fuel cell car by 2015—I suspect primarily to improve public relations. Without any serious hydrogen infrastructure in the United States (which has only 56 stations nationwide, including 9 in New York and 23 in California) or in Japan (12 stations in the entire country), there will be no incentive (or need) for Toyota to mass-produce the car. Like the Tesla Roadster, it will be left behind in history.

Will fuel cells be used in stationary applications, to replace power plants? The direct competition is combustion turbine technology. A stationary natural-gas power plant, operating in the "combined cycle" mode (with waste heat from the first turbine driving a second one) can achieve over 50% energy-to-electric efficiency, and it can do so with a capital cost investment that is 7 times lower.

19

CLEAN COAL

COAL is the filthiest fuel we have. There are many reasons.

- Coal is primarily carbon, so for each kilowatt-hour produced, coal produces twice as much carbon dioxide as does natural gas.[58]
- Burning low-grade coal makes sulfur dioxide. Sulfur dioxide mixes with water in the atmosphere to become sulfuric acid, and when sulfuric acid precipitates, it's called *acid rain*. Acid rain destroys forests (Figure III.25), damages marble monuments, and changes the acidity of nearby lakes.
- Coal is the source of the mercury that contaminates our fish.
- Coal power plants create fly ash and those small black carbon particles that soil your hair, accumulate on your windowsill, and are called "black carbon" when they melt Greenland ice.
- Coal is responsible for the overwhelming air pollution in many Chinese cities and for the resulting severe respiratory problems, amounting to a national emergency. During the Olympics, the Chinese government turned off all the coal plants within miles of Beijing, lest the world actually see on television the unbreathable air that normally chokes that city.

Figure III.25. Forest damage attributed to acid rain caused by nearby coal plants.

Why would anyone like coal? The obvious answer: coal is dirt cheap. If you live close to a coal mine (Wyoming, Idaho, Kentucky), you can buy electricity for 6¢–7¢ per kilowatt-hour (I pay 15¢ in California). Build a new conventional coal plant and you could sell the electricity for 8.6¢ per kilowatt-hour, according to the Energy Information Administration. (California doesn't allow new coal because of greenhouse gas concerns.)

Because of the local contamination and its serious health effects, China is desperately working to clean up the exhaust from its coal plants. The particulates can be removed with electrostatic precipitators—basically, metal plates charged to high voltage to attract the particles and remove them from the gas. Sulfur dioxide can be washed out by spraying the gas with various kinds of soap and other chemicals (sodium hydroxide, lime, sodium sulfite, ammonia). These approaches are expensive, and they drive up the cost by 1¢–2¢ per kilowatt-hour. That's reasonable enough, and the health costs of not doing it are so high, that I fully expect China to add scrubbers to most of its plants over the next few decades.

Does scrubbing make coal a clean energy source? In the traditional sense, yes. Install scrubbers and you will no longer see black smoke billowing from tall smokestacks. Coal operations that do this have been traditionally called "clean coal" plants. Nowadays, many people don't consider traditional clean coal to be clean enough. A coal plant that produces a gigawatt of electricity emits a ton of carbon dioxide every 2 seconds.[59]

Coal haters have their mantra: "Clean coal is an oxymoron." Look up this phrase on Google or Bing and see the long list of politicians and commentators who have used that phrase recently. I'd guess that millions of Americans have learned the meaning of the word *oxymoron* from the widespread use of this epithet. It refers to the fact that clean coal produces enormous amounts of greenhouse gases. China is adding one new gigawatt coal plant every week. That means that next week the carbon dioxide emissions of China will be larger than they are this week by a ton every two seconds.

Yet President Obama has publicly and repeatedly supported clean coal. What's going on? President Obama is not ignorant of these issues; it is mostly the oxymoron people who are confused. When Obama refers to "clean coal" he doesn't mean the "old clean coal"[60] of smokestack scrubbers, but the "new clean coal" of carbon sequestration.

Carbon sequestration refers to the goal of preventing the carbon dioxide generated by coal burning from getting into the atmosphere. A more accurate term would be "carbon *dioxide* sequestration." Other equivalent terms are "carbon dioxide capture and sequestration"—referred to as CCS, although some people will claim that CCS stands for "carbon capture and storage." Don't worry about the terminology; the real issues are twofold: can carbon dioxide be buried at an affordable price, and will it continue to stay securely buried over the long term?

It is not easy to separate carbon dioxide at the smokestack. The best approaches to CCS require redesigning the entire coal plant, improving efficiency (to maximize the energy per ton), and capturing the carbon in the plant itself, not in the stack. The premier US project to do this was called FutureGen. In the first stage of Future-Gen, energy was used to separate oxygen from air so that the fuel could be burned with pure oxygen. Nitrogen was left behind. It turns out that using pure oxygen instead of air helps both the efficiency of the turbine and the ease of later separation of the carbon dioxide. Gasify the coal by reacting it with hot water, producing hydrogen gas and carbon dioxide. These are easily separated. The hydrogen is burned in a turbine, and the waste heat is used to drive a second steam turbine. Use of two turbines gives rise to the term *combined cycle* to describe this system; FutureGen was called an *IGCC*, standing for "integrated gasification combined cycle" power plant. The carbon dioxide output is compressed and pumped underground. There have been many suggestions for storage locations, and the IPCC wrote a special report on this subject. One option is to pump the gas into old oil wells, thus helping with enhanced oil recovery. Already, more than 10,000 wells in Texas are using this method. It's a great solution because it's sequestration that is profitable.

Carbon dioxide could also be stored in empty coal mines and in depleted oil and gas wells. My favorite locations are underground reservoirs in rock layers currently containing salt brines. These have no commercial value, are found all over the world, and appear to have the required impermeability to keep the carbon down for hundreds of thousands of years or longer. Over 11 million tons of CO_2 have already been sequestered in such brines in the Sleipner field off shore in Norway.

In 2009, FutureGen was canceled. The stated reason was that the project was far over budget. At that time I talked to a strong supporter of clean coal (I'll keep his name secret), and he expressed relief; he was glad it was canceled. His concern was that Future-Gen was being done at such a grandiose level that if and when it was completed, it would work but it would teach the wrong lesson: that capture and sequestration was a terribly expensive thing to do. FutureGen's success, in his mind, would convince the world that CCS was unaffordable.

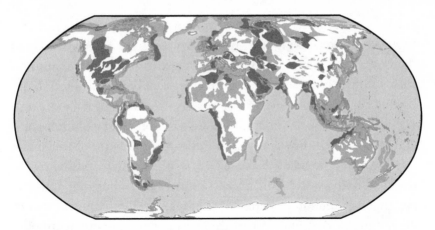

Figure III.26. Potential locations for carbon dioxide sequestration, according to the IPCC.

As the IPCC map in Figure III.26 illustrates, possible sequestration sites can be found all over the world. Will all the brine sites have the right kinds of impermeable caps to keep the carbon dioxide down indefinitely? Unfortunately, we don't know for sure, so we can't be certain that this method will work. However, it looks good on paper, and it's worth trying. If carbon dioxide capture and sequestration is going to fail, we'll learn soon, from initial leaks.

Clean coal is deeply susceptible to skepticism bias. If you read about it, you'll hear the constant criticism that it is "unproven." But what alternative-energy sources have been proven? Is cheap solar proven? Is low-cost battery technology proven? Today, the only financially possible solution for China to reduce its carbon dioxide growth (with its 1 new gigawatt of coal power every week) is conversion to natural gas; but even that produces carbon dioxide. If you are truly worried about global warming, then carbon capture and storage may be the only long-term hope—at least if China continues its spectacular growth in energy use. You can be optimistic about solar, but remember, China added only ⅛ of a gigawatt (average power) of solar last year. It is truly biased to apply skepticism to one technology and optimism to another.

Sequestration will not be cheap. The carbon dioxide must be compressed to be injected into the deep brines, and that takes energy and dollars. The US Energy Information Administration estimates that sequestration will add an additional 3¢ per kilowatt-hour to the cost of electricity. Many people think it will be more, but let's stick with the EIA estimate. Considering that new coal plants in the United States deliver electricity at a cost to the consumer of about 10¢ per kilowatt-hour, this seems like a reasonable premium to pay.

The challenge, however, is not in the United States. Most of the future carbon dioxide will come from coal in China and the rest of the developing world. Will those countries be willing—or able—to pay that 3¢ premium? China now ranks second in the world at producing electricity, and first in the world at producing greenhouse gases. Look back at Figure I.16, which plots the United States' and China's changes in CO_2 emissions through 2010. In 2010, Chinese emissions were 70% greater than those of the United States, and by now they may be double ours.

What would it cost to sequester all of that carbon dioxide? The IPCC concluded that it would raise the cost of the energy from coal by 50%–100%. That's more than the EIA estimate of 3¢ per kilowatt-hour. But let's use the lower estimate. The total Chinese electric production from coal for 2011 was greater than 4 trillion kilowatt-hours; at 3¢ per kilowatt-hour, the cost for sequestration would be $120 billion *per year*. Moreover, Chinese electric use has been growing faster than 10% per year, so we might expect these yearly costs to also rise at that same rate.

Can China afford the cost? It is a poor country and might be reluctant to spend that much money on pollution that doesn't remain in China (which occupies less than 2% of the Earth's surface area). A sum of $120 billion per year spent elsewhere could do much to help the Chinese reduce poverty, improve health and education, clean the local environment, and stimulate continued growth. If you were premier of China instead of US president, what would you decide? Would you devote these funds to reducing greenhouse gases, instead of using them for purposes that give immediate benefit to your own people?

When you do become president, consider the fact that most of the greenhouse gas pollution will be coming from the developing world. The best way to spend the $120 billion each year would not be for advanced industry in the United States—for advanced power grid or advanced automobile technology or advanced solar and wind power and nuclear power. No, the optimum way to spend any amount of money for this purpose might be to send it to China to help the Chinese sequester the emissions from the new gigawatt coal plants they're building every week. I am a physicist, not a politician, but it is my guess that if you propose that approach, you will not help your reelection chances.

A more realistic solution would be to encourage China to convert from coal to natural gas. It would reduce emissions by only 50%, but that is substantial, and (if our attribution to humans is right), it would certainly slow the onset of global warming. That, in turn, could give us time to further develop other carbon-free energy sources.

IV

WHAT IS ENERGY?
—AN OPTIONAL CHAPTER

We must conserve energy.   —*President Jimmy Carter*

Energy is conserved.   —*Any physics professor*

Energy is like love: you don't have to understand it in order
to get involved with it. But unlike love, there is some chance
you actually can understand energy.   —*Anonymous*

THIS *is a part of the book that you really don't need to read. Feel free to
skip it. Or even better, read though it casually, without feeling you have to
learn it. Energy is a very useful tool, even if you don't deeply understand
its inner nature. Energy can be commonplace and pedestrian (you can
measure it in food calories), or it can be abstract and ethereal (as when
physicists say it is the conjugate variable to time).*

The Properties of Energy

ENERGY IN FOODS, FUELS, AND OBJECTS

Let's start with the pedestrian. Table IV.1 shows the available energy
in a pound of various foods and fuels and objects. Look at the last
column, the energy compared to that of TNT. Do any of the entries
surprise you?

Table IV.1. Energy per pound for various objects and substances.[1]

	Food calories	Kilowatt-hours	Compared to TNT
Bullet (1,000 ft/s)	4.5	0.005	0.015
Auto battery	14	0.016	0.046
Computer battery	45	0.053	0.15
Alkaline battery	68	0.079	0.23
TNT	295	0.343	1
High explosive (PETN)	454	0.528	1.5
Chocolate chip cookies	2,269	2.6	7.7
Coal	2,723	3.2	9.2
Butter	3,176	3.7	11
Ethanol	2,723	3.2	9
Gasoline	4,538	5.3	15
Natural gas (methane)	5,899	6.9	20
Hydrogen	11,798	14	40
Asteroid (30 km/s)	48,435	57	165
U-235	9 billion	11 million	32 million

My favorite surprise in this table (it surprised me when I first calculated it) is that chocolate chip cookies deliver 7.7 times more
energy than an equal weight of TNT. Amazingly, gasoline has 15
times the energy of TNT! With such low energy per pound, why do
we ever use TNT, rather than gasoline or cookies? The answer: TNT
can release its energy much faster, so it delivers high power (energy
per second); high power can create large force, and that's what is
needed to crack rock and concrete. I use this example in my class to
illustrate the difference between energy and power; power is the rate
of energy delivery. If you want a very big explosion but are limited
by the weight you can carry, use gasoline instead of high explosives;
that's what our military did with the fuel-air explosives it dropped
over Afghanistan.

Is Energy a Thing?

I still haven't answered the question posed in the title of this chapter: What *is* energy? Let me put that off a little longer and take a

moment to remind you of several properties that are truly mysterious but are so common that you might not realize how mysterious they are. When you hit a baseball with a bat, energy is transferred. But no *thing* is transferred; the bat gives up a property (high velocity) to the ball, but it does not transfer an object or material. The situation is similar with a sound wave: the individual molecules in air don't move very far when sound passes, and they wind up in the same location where they started. They moved for a short while and what they passed on was not their mass or their identity, but their pattern of motion. Is this pattern energy?

Certainly a pattern can hold energy, but energy is more complicated. Think about the subcategory of energy that we call heat. *Heat* is the random motion of molecules; at room temperatures, the molecules are vibrating with an average instantaneous velocity of about 770 miles per hour. Some are moving at half that speed; some at twice; 770 is the average. Thus, for a bullet moving at 770 miles per hour, many of the molecules are actually moving backward— those whose random motion is going opposite to the bullet's forward motion. In fact, there is as much energy stored in this random motion as there is in the forward motion of the bullet. The energy in that coherent velocity is what does the damage when the bullet hits; the equal amount of energy in the heat just keeps the bullet warm.

Suppose you have one bullet sitting at rest and another bullet moving at 770 miles per hour. Compare the heat energy of the resting bullet to the kinetic energy of the moving one. They are equal. Unfortunately, most of the heat energy can't be extracted for useful work, because it is *disorganized*. The molecules move in random directions. (The concept of disorganization is formalized in physics in the theory of *entropy*.) There is a general physics theorem about disorganized energy: the fraction that can be converted to useful organized energy—the efficiency—is given by a simple equation called the *Carnot equation*:

$$\text{Efficiency} = \left(1 - \frac{\text{Cool temperature}}{\text{Hot temperature}}\right) \times 100\%$$

When you use this equation, the hot temperature is typically the temperature of the combusted fuel, or of the water heated by concentrated sunlight. The cool temperature is typically the temperature of the exhaust gas, or of the cooling tower. The only catch is that for this equation the temperature has to be expressed not in degrees Fahrenheit or Celsius, but in degrees absolute. To do that, either add 459 to the Fahrenheit numbers or add 273 to the Celsius numbers.

The Carnot equation helps explain the limited efficiency of any motor that runs on heat, such as a turbine or internal combustion engine. In the discussion of geothermal energy in Chapter 15, I used this equation to show that water heated by rock by 30°C can deliver only 9% of its energy. Now consider water heated to boiling: 212°F, a temperature relevant not only for geysers but also for boiling-water nuclear reactors. Assume we extract energy and the water cools to room temperature, 65°F. The hot temperature is 212 + 459 = 671. The cold temperature is 65 + 459 = 524. So the maximum efficiency of energy extraction is

$$\text{Efficiency} = \left(1 - \frac{524}{671}\right) \times 100\% = 22\%$$

This means that only 22% of the heat in boiling water can be extracted to make electricity or other useful energy.

When President Carter asked us to conserve energy, he was referring to *useful* energy. Electric energy is very useful; it can be converted efficiently to mechanical motion or (if it's what you want) to heat. But once it is heat, only part of it can be transferred back—the percentage given by the Carnot efficiency equation.

The Meaning of Energy

What is energy? There is a short answer, but it is so abstract that we normally take 4 years to prepare physics majors to learn it. I'll condense those years of initiation into a few paragraphs. As you travel through these, you'll get a sense of the intellectual transformation experienced by the maturing physicist.

ENERGY AS TAUGHT TO HIGH SCHOOL STUDENTS AND COLLEGE FRESHMEN

Introductory energy is the least interesting kind. Please don't get so bored with this section that you don't bother reading on; the deeper, more abstract, and exciting definitions of energy are coming up. To the beginner, *energy* is defined as the ability to do work. Work = force × distance. Force is something that accelerates mass: Force = mass × acceleration; that is, $F = ma$.

Virtually no freshman physics major really understands these equations, for many reasons. The word *ability* is undefined, and perhaps undefinable. The equation $F = ma$ is not really a law but rather just a definition of the term *mass*. (The *law* is that mass, the m in the equation, is a constant independent of velocity.) It doesn't matter; as with this book, the first step is to learn *about* energy, and how to handle the language and the equations. A student who can learn the rules and how to use them can earn an A in the course without truly understanding anything.

ENERGY AS TAUGHT TO SOPHOMORES

Now it gets a little more interesting. Energy can be converted to mass and vice versa. This fact is often expressed by Einstein's equation:

$$E = mc^2$$

Many people are confused by this equation because it works only if you plug in the numbers in the right way. If you express the mass m in kilograms, and the speed of light c in meters per second ($c = 3 \times 10^8$ meters per second), the equation gives the energy in joules. If you want to convert to kilowatt-hours, divide by 3,600,000. And the mass m is no longer a constant. If the object has a velocity, the mass increases.[2]

Sophomores also learn the Planck law—that the energy E in quantum physics is related to the quantum frequency f by $E = hf$, where h is a number we call Planck's constant. Although they learn this simple law, most sophomores don't understand what it means just yet. Frequency is measured in cycles per second. The Planck equation seems to be hinting at a deep relationship between energy and time. It is.

Energy as Taught to Juniors

Energy and mass are actually the same thing. If energy is present, not only does it have mass—in fact, it *is* mass. Energy and mass are not only equal, they are equivalent; that's the real meaning of $E = mc^2$. We see this in atoms. Some kinds of energy are negative, such as the *binding energy* that holds the electron to the nucleus; most amazing, the presence of this negative binding energy lowers the mass of the atom! It's as if negative energy contributes negative mass. Indeed, even the gravitational force—something that originates in mass—from that atom is lower because of that negative mass.

Juniors also get a second glimpse at the deeper meaning of energy—its relation to time. (The first glimpse came from the Planck law.) The only objects that have a unique energy value are those whose quantum mechanical wave function oscillates with time exactly like a sine wave. Again, the energy is given by $E = hf$. If the wave function oscillates differently (like a square wave, or a sum of two beating sine waves), then the energy is "uncertain"—meaning that if you try to measure it, you will get one of a list of possible values, and not one definite predictable value. This is the heart of the Heisenberg uncertainty principle and the Planck equation. And there is another hint at the connection between energy and time: the uncertainty in energy, multiplied by the interval in time used to measure the energy, is always greater than Planck's constant. Energy and time seem to be linked.

Energy as Taught to Seniors and Graduates

The most fascinating, precise, and (for the physicist) practical definition of energy is the most abstract one—too abstract to even be discussed in the first few years of a physics education. It is based on the observation that the true equations of physics, such as $E = mc^2$, will be as true tomorrow as they are today.[3] That's a hypothesis that most people take for granted, although some people continue to test it; if a deviation is found, it will mean a certain Nobel Prize. In the jargon of physics, the fact that the equations don't change is called *time invariance*. It doesn't mean that things in physics don't change; as an object moves, its position varies with time, its velocity varies with time, lots of things in the physical world change with time— but *not* the equations that describe that motion. Next year we will still teach that $E = mc^2$, because it will still be true.

Time invariance sounds trivial, but when you express it mathematically, you can derive an astonishing conclusion: you can prove that energy is conserved. The proof was discovered by Emmy Noether (pronounced NER-ter), a contemporary of Einstein, who called her one of the most "significant" and "creative" mathematicians of all time (Figure IV.I). Like Einstein. she fled Nazi Germany and came to live in the United States.

Following the procedure outlined by Noether, starting with the equations of physics, you can always find a combination of your variables (position, speed, and so on) that will not change with time. When you apply this method in the simple cases (classical physics,

Figure IV.I. Emmy Noether, who discovered the link between time and energy.

with force and mass and acceleration), the quantity that doesn't change with time turns out to be the sum of kinetic and potential energy—in other words, the classical energy of the system.

Big deal. We already knew that energy was conserved. But now there is a fascinating philosophical link. There is a *reason* why energy is conserved! It's because of time invariance.

And there is an even more important result: the procedure works even when we apply the method to the much more complex equations of modern physics. Imagine the following question: In the theory of relativity, what is it that is conserved? Is it energy, or energy + mass energy? Or something else? And what about chemical energy? And potential energy? How do we calculate the energy of an electric field? What about quantum fields, such as those that hold the nucleus together? Should they be included? Question after question with no intuitive answer.

Today, when such questions arise, physicists apply the procedure outlined by Noether and get the unambiguous answer. When Einstein applied the method to his relativistic methods of motion, he derived the new energy, one that contained mass energy, mc^2. When we apply the Noether method to quantum physics, we come up with terms that describe the quantum energy.

Does this mean that the "old energy" was not conserved? Yes it does; if we have improved equations, then not only are the predicted motions of particles different, but also the things we thought were conserved aren't. Classical energy is no longer constant; you *must* include the mass energy—and the energy of the quantum fields. By tradition, we call the conserved quantity the "energy" of the system. So although energy itself doesn't change with time, as we dig and uncover the deeper equations of physics, our *definition* of energy does change with time.

The Beauty of Energy

If you're still with me, let me go a little further. Think about this question: Do the same physics equations that work in New York City also work in Berkeley? Of course. Actually, that observation is not trivial; it has extremely important consequences. We say that the equations don't depend on location. We may have different masses, or different electric voltages—but those are the variables. The key question is whether the equations that describe the physics of the behavior of objects and fields is different in different places.

The equations that we have in physics today—all of the ones that are part of the standard physics, the ones that have been veri-fied experimentally—have the property that they work everywhere. Some people think this is amazing enough that they spend their careers looking for exceptions. They look at things that are very far away, such as distant galaxies or quasars, hoping to find that the laws of physics are a little bit different. So far, no such luck. No quick Nobel Prize. Not yet.

Now to the remarkable consequence. The same Noether math that worked for equations that don't change with time also works for equations that don't change with location. If we use Noether's method, we can find a combination of the variables (mass, position, velocity, force) that does not change with time. When we apply this procedure to the classical physics invented by Newton, we get a quantity that is equal to the mass multiplied by the velocity; that is, we get the classical *momentum*. Momentum is conserved, and now we know why. It's because the equations of physics are *invariant in space*.

The same procedure can be used in the theory of relativity and in quantum physics, and in the combination called relativistic quan-tum mechanics. The combination that doesn't change with time is a little different, but we still call it momentum. It contains relativis-tic terms—as well as the electric and magnetic fields, and quantum effects—but by tradition we still call it the momentum.[4]

Perhaps I should add a few more comments here about quantum physics and the Heisenberg uncertainty principle. According to quantum physics, even though we can define them, the energy and momentum of a *part* of a system are often uncertain. We may not be able to determine the energy of a particular electron or proton, but the principle does not have a similar uncertainty for the total energy of a system. The whole collection together can shift energy back and forth between its various parts, but the total energy is fixed; energy is conserved.

Let me go even further. In relativity theory, physicists see space and time as deeply intertwined; we refer to the combination as *space-time*. The invariance of physics in time leads to energy conservation.[5] The invariance of physics in space leads to momentum conservation. If we put the two together, the invariance of physics in space-time leads to the conservation of a quantity called energy-momentum. Energy and momentum are seen, by physicists, as two aspects of the same thing. From this point of view, physicists will tell you that energy is the fourth component of the four-dimensional energy-momentum vector. If the three components of momentum are labeled p_x, p_y, p_z, then the energy-momentum vector is (p_x, p_y, p_z, E). Different physicists count the four components differently. Some think energy is so important that they like to put it first. They then call energy the zeroth component of the vector instead of the fourth: (E, p_x, p_y, p_z).

Physicists tend to get almost religious when they think about the advanced definition of energy—the quantity conserved because of the time invariance of the equations of physics. It sounds simple and yet it is very deep. It relates two concepts that would normally appear totally unrelated: energy and time. They call energy and time "conjugate variables." Their connection provides an insight that physicists feel goes beyond engineering and borders on the spiritual. The profound relationships such as that between energy and time—and between momentum and space—are what physicists think of as the "beauty" of physics.

You don't have to agree that this represents beauty. You may find a rainbow or the eyes of a child to be far more thrilling. But at least now you know what the physicist is referring to.

V

ADVICE FOR
FUTURE PRESIDENTS

THE ROLE *of a science advisor should not be to advise but rather to inform and educate so that the president knows and understands enough to be able to make the right decisions. When you are elected, you'll have to balance many things, including diplomacy, economics, justice, and politics—things that scientists like me have not mastered. Any advice I give is based on my narrow technological understanding of the world, and for that reason it has limited value to you.*

But just in case you're curious about my opinions anyway . . .

THE TRUE energy crisis in the United States, and in much of the rest of the world, derives predominantly from two issues: energy security and global warming. The security problem comes not from an energy shortage (we have plenty), but from an oil shortage—more precisely, from the growing gap between domestic petroleum production rate and the demand for gasoline, diesel, and jet fuel.

The global-warming problem derives primarily from rapidly growing coal use in the developing world.

Petroleum scarcity has forced us to enormous imports, resulting not only in military insecurity (could we fight a war if oil imports were cut?) but also in a huge balance-of-trade deficit that threatens our economic well-being. Immediate action to relieve this crisis requires rapidly exploring our great shale gas and oil reserves, building an infrastructure to handle the distribution, and creating a strong synfuel capability.

To reduce the global-warming threat from carbon dioxide emissions, we must recognize that this gas will come primarily from the developing world. Simply spending money to set an example that poor nations can't afford to follow is not a solution. The only plausible method that holds up to scrutiny and that may be economically viable is large-scale conversion from coal to natural gas.

The era of natural gas is upon us. It will outcompete most alternatives, although wind, solar, and nuclear have a fighting chance. Energy from gasoline and oil costs 2.5–5 times as much as from natural gas. Synfuel, made from natural gas or coal, will be a key part of our energy future. Its existence should keep the long-term cost of gasoline well below $3.50 per gallon, but that limit will kick in only when we have plenty of synfuel factories operating. Natural gas, synfuel, and shale oil are the three technologies that will predictably have the biggest impact in reducing our balance-of-trade deficits.

Following is my categorization of the importance of the major energy technologies.

Technologies that will be an important part of our energy future:
- Energy productivity (efficiency and conservation)
- Hybrid and other autos with improved mileage
- Shale gas (for coal replacement, autos, synfuels)
- Synfuel (gas to liquid and coal to liquid)
- Shale oil
- Smart grid

Technologies that have breakout potential:

- Photovoltaics (PVs)
- Wind (and an improved grid to deliver it)
- Nuclear power (both old and new generations)
- Batteries (to back up PVs and wind)
- Biofuel (particularly grasses such as *Miscanthus*)
- Fuel cells (particularly methane-based)
- Flywheels

Technologies that are least likely to solve our problems:

- Hydrogen economy
- All-electric autos and plug-in hybrids
- Corn to ethanol
- Solar thermal
- Geothermal
- Wave and tidal power
- Methane hydrates
- Algae biofuel

How can you, as president, encourage or enhance the right technologies? For some, vigorous research programs are appropriate.
For others, careful regulation is needed (such as making sure that
fracking is done in a clean and environmentally conscious way). For
still others, it is only necessary that overly restrictive regulations be
removed; for example, some of the nuclear power regulations—such
as requirements for emergency core cooling systems—are based on
early-generation designs but aren't needed or appropriate for later-
generation models. Let's review some of the policy implications of
these technologies.

Energy Technology Policy

ENERGY PRODUCTIVITY

The cheapest form of energy is energy not used—*invisible energy, negawatts*. You can achieve energy savings through conservation and efficiency, but beware of those terms. Conservation was given a bad name by President Carter, and efficiency is often wrongly associated with low profits by businessmen. Moreover, efficiency doesn't save energy if it encourages more use. For those reasons I use the term *energy productivity*, and you might consider using that too. There are enormous gains to be made, as the McKinsey chart shows (Figure II.16, in Chapter 7). Utilities can be engaged in energy productivity through the right kind of incentives; one of the most successful ones so far is the obscurely named "decoupling plus," a system that allows utilities to profit by selling productivity instead of energy.

Energy productivity is key, and I'll return to it later in this section.

NATURAL GAS

We have a natural-gas windfall. If we develop it properly, it will help enormously over the next few decades—not only in getting us past our petroleum crisis but also in reducing carbon dioxide emissions. We need to encourage an infrastructure for natural-gas service stations. At the same time, we need strong laws that prevent the local pollution that can result from fracking.

Natural-gas use will grow rapidly, not just in the United States but around the world. This fuel is going to be so important that you might consider launching a nationwide program, called something like *The Natural Gas Economy*, that recognizes the value of the new gas source and develops a coherent policy and infrastructure to encourage its exploitation.

SHALE OIL

Just a few years ago, many people concerned about energy completely underestimated the looming importance of shale gas. Now they may be doing the same with shale oil. This potentially disruptive (in the good sense) technology is just beginning to break out, and I predict that in a few years it will be enormously important. By the end of this decade, 25% of our oil may come from this source. Shale oil could solve our energy security problems and reduce our balance-of-trade deficits. It is conceivable that it could actually eliminate the deficit, by making the United States once again an *exporter* of oil. Experts estimate that much of the shale oil can be recovered at a cost as low as $30 per barrel. If that turns out to be true, then shale oil will offer stiff competition to synfuel, although natural gas will still offer greater energy per dollar.

Petroleum is used primarily for transportation, and any new oil source will be opposed by those who see the automobile as evil. But shale oil will be hard to resist, from both economic and energy security reasons. The best approach to the environmental concerns might be to continue strengthening the CAFE standards for automobile efficiency (these are the restrictions placed on the "Corporate Average Fuel Economy" of the carmakers), pushing toward the achievable goal of 100-mpg autos.

SYNFUEL

A vigorous program for synfuel is important not only for our security (to provide emergency fuel for our military) but also for our economy, to reduce our deficit of payments. The main criticism of synfuel will be that it is not green and not sustainable. But keep in mind that the contribution of US automobiles to global warming over the past 50 years was only $1/40°C$, and that with reasonable CAFE standards, the US auto contribution can be kept low in the future. As for "sustainable," we have enough natural gas to last for many decades, and enough coal to last for a hundred years or more, and after that we'll probably be running our automobiles on infinitely rechargeable batteries or nukes or antimatter or something we can't even imagine now.

HYBRIDS AND PLUG-IN HYBRIDS

The all-electric automobile is *not* going to be a major contributor to the transportation energy future of the United States unless there's an enormous breakthrough in battery technology—and that's unlikely. All-electric autos should not be subsidized or even encouraged. But the development of batteries, which are important for many things besides electric autos, deserves continued support. Static batteries for energy storage are very different from those needed for autos. On the other hand, in the developing world, where automobile ranges of more than a hundred miles are not demanded by consumers, lead-acid batteries could compete with gasoline engines.

A push toward electric automobiles in the United States ignores the high cost of battery replacement. I suspect that the all-electric enthusiasm is driven by the hope that as the market for batteries grows, their cost will plummet. But battery technology, particularly for autos, is very challenging. Batteries are unlikely to drop rapidly in price, and they have a very long way to go to reach cost parity with fossil fuel.

The enthusiasm for electric automobiles is also driven by an exaggerated belief that they will be essential in keeping down carbon dioxide emissions. Don't forget that an electric auto recharged from a coal power plant puts more CO_2 into the atmosphere than does a gasoline auto.

On the other hand, hybrid automobiles are destined to become a much more widespread technology. They, combined with use of lightweight but strong materials, will prove to be one of the most effective means for improving automobile mileage. I predict that most of us will be driving hybrids within the next decade or two.

NUCLEAR

Encourage a vigorous nuclear program. It may take you, when you are president, to convince the American people of the importance of this technology to the future of energy in the United States and the world. You can do this by knowing the facts, and by explaining that nuclear can be a safe and viable component of our energy future. You need to make sure the public knows that so far there have been no deaths from radiation leakage at Fukushima, and that the expected number of eventual deaths from Fukushima-induced cancer is likely to be less than 100—tiny compared to the 15,000 killed by the tsunami. The horror of the tsunami was the tsunami itself, not the fact that it destroyed a nuclear plant. Consider using the "Denver dose" as your standard: if any radiation level is lower than that in Denver, we will not concern ourselves with it. Of course, you can judge better than I can the political fallout of this approach.

Nuclear is already safe and clean, and we need to set an example for the rest of the world that reflects that fact. Nuclear will also be very important in reducing emissions of the developing world. Some of the new third- and fourth-generation nuclear power plants are well suited to poorer nations. With a thoughtful policy, the United States could become a major manufacturer and supplier of such power plants.

It was a terrible mistake to close the nuclear waste facility in Yucca Mountain, Nevada. Storing spent nuclear fuel at that location is plenty safe, and enormously safer than keeping the waste on site at the reactors (as had been done at Fukushima). The Yucca Mountain facility needs to be reopened and expanded. There's lots of space there for more waste, but new tunnels are required. We should also build more storage facilities. Reopening Yucca Mountain would also provide important leadership. Engineers around the world have difficulty convincing their governments that nuclear waste storage is a solved problem when their prime ministers say, "Then why does the United States believe differently?"

Key Considerations

GLOBAL WARMING AND CHINA

Global warming, although real and caused largely (and maybe 100%) by humans, can be controlled only if we find inexpensive (or better, profitable) methods to reduce greenhouse emissions in China and the developing world. Beware of feel-good measures (like electric cars) that will have tiny ($\frac{1}{40}$°C) effects on global warming but can mislead people into thinking they're addressing the concern.

Your most difficult energy decision may be what to do about China. Its greenhouse emissions back in 2010 were already 70% larger than those of the United States, and they may have grown to double the US level by now.

Perhaps the best way to reduce China's future carbon dioxide emissions would be to facilitate its switch from coal to shale gas. There is very little in the way of proprietary technology in horizontal drilling and fracking, but that doesn't mean the method is easily adopted. What shale gas recovery really demands is a mass of specialized equipment and highly trained staff. In May 2012, Marlan Downey and I proposed a practical solution in an article, "Fracking Global Warming": we invite 100 Chinese engineers to come to the United States for a year, to be trained in shale gas technology by close observation of US operations. On their return, their new expertise would facilitate and expedite a shale and coal gas boom in China. Rapid development of these resources would enable China to convert from coal, which poses severe health problems for the Chinese people, to natural gas, which produces only half the greenhouse emissions, and none of the mercury or sulfur pollution that harms the local population. This proposal should be supported even by those who don't believe that humans cause warming. A switch from coal to natural gas is such a great benefit for the health of the Chinese people that it is worth doing for humanitarian reasons alone.

Some people will oppose fracking because of the dangers of local pollution. But of all the technical problems from our energy challenges, this is certainly the easiest one to address. Make strong legislation that requires all waste from fracking operations to be clean—drinkable by humans. Make the fines big; fracking is profitable, and the cleanup is affordable.

China also needs to produce as much solar and wind power as

possible. Here's a political conundrum for you. Maybe the best way to achieve this goal is to encourage the Chinese solar and wind industries. If China is willing to subsidize those technologies, that's wonderful . . . or is it? In the world market, the Chinese compete directly with our own solar and wind industries, which the United States subsidizes with loan guarantees. Already US solar companies are being driven out of business by the cheaper Chinese cells. What's the solution? How do you balance the value of a vigorous Chinese industry with the value of a vigorous US industry? Whatever the answer is, it's beyond the ability of a science advisor to advise. Good luck with this one.

Because Chinese consumers are not yet addicted to cars with ranges in the hundreds of miles, there is a potential market in China for electric automobiles based on lead-acid batteries. Some people worry that lead is polluting, but in poorer countries lead is too valuable to dump; rather, the batteries that contain it are reconditioned and recycled.

More on Energy Productivity

Energy productivity is win-win-win: it saves money and reduces imports while stimulating the economy. There are huge gains to be made. Automobiles can be made much more efficient, but market forces are not likely to accomplish this improvement on their own, in part because of the *paradox of the commons*. If you're not familiar with this famous paradox, it is worth looking up. When benefits and resources are shared (as in the "commons" grazing field in old British towns), the greatest benefit will come if they are used proportionately; but that situation is unstable because one individual can profit excessively by being greedy. With automobile efficiency, everyone would benefit if, for example, every auto were lighter. But then one person, getting a heavier car, would be safer, while endangering others. The solution is a communal law, the CAFE standard, to ensure that the required economies are widely shared.

When I bought my first car, decades ago, typical mileage was 14 mpg, and the Volkswagen "Beetle," by being super small and slow to accelerate (and also having a superhot engine that produced lots of nitrous oxides), got 32 miles per gallon. That seemed spectacular at the time (even with gas at 29¢ per gallon). Soon the CAFE average in the United States will be 35 mpg for the entire fleet, which includes heavy, large, high-horsepower cars. Yes, we've done what once seemed impossible, and we can still do a lot better. You need to inform the public that fiber composite bodies are as safe as or safer than heavy metal bodies; that cool paints on autos can save gasoline used to run the air conditioner; that improved CAFE standards can indeed be achieved without reducing performance or the consumers' perceived comfort level.

Other aspects of energy productivity can be improved enormously. The McKinsey chart (Figure II.16, in Chapter 7) shows numerous ways that saving energy can return a profit. Investments in technologies that reduce energy use—cool or white roofs, better insulation in homes, more efficient appliances, compact fluorescent or LED lighting—can all yield returns on the investment better than those of Madoff's Ponzi scheme. And what's more, the returns are tax-free. Legally.

The US public needs to understand that conservation does not mean a reduced standard of living. With proper conservation measures, you can turn up the thermostat of your home if you want to. You will make money on your energy productivity investments while not even noticing a change.

It may be politically wise to avoid conservation measures that will be perceived by the public as interfering with their lifestyle. An example is forced conversion from tungsten filament lightbulbs to fluorescents. Unfortunately, many people will get the wrong kind of bulbs and see themselves looking pale in their bathroom mirrors; the result could be a great reaction against further conservation measures. On the other hand, if you subsidize compact fluorescent bulbs to encourage use, you may save money by not needing to build as many power plants. If you do this, make sure you subsidize the bulbs that have a warmer color temperature.

ELECTRIC POWER GRID AND VENTURE CAPITALISM

If you feel you must spend additional stimulus money, then spend it on improvement of the electric grid. Such an infrastructure can make wind and solar far more accessible and profitable. You can also reduce the 7% energy loss that our current grid suffers.

Beware of trying to pick winners. Venture capitalism is a very difficult business, and venture capitalists typically expect three out of four of their investments to go under. Government probably can't sell that kind of failure rate to the public, and even 25% success requires enormous skill. If you put amateurs or politicians or academics in charge, they are likely to choose the wrong companies; they'll underfund the better prospects and overfund the eventual losers. Venture capitalists succeed, in part, because their livelihood depends on succeeding. Theirs is a finely honed skill, and even many venture capital firms fail and go out of business.

SUBSIDIES

Subsidies make sense only if they inspire rapid competitive development. The plunging cost of solar is due to a remarkable combination of subsidies and competition. At the same time, the survival of large-scale solar thermal, a technology that in my opinion has little long-term prospects, was made possible by the same subsidies.

ENERGY CATASTROPHES

Energy catastrophes are a difficult challenge. You need to be careful not to exacerbate their negative impact by frightening the public, yet you will face a political challenge if people think you're downplaying real dangers. You will be tempted to call every accident a disaster; doing that has proved the safest political bet for previous presidents. Take that easy approach, and if you fail to fix the problem you will be forgiven; if you overcome the problem (or don't need to because it wasn't really so bad in the first place) you will be considered a hero. Exaggeration, however, is a dangerous path, because the public doesn't like to discover that it has been fooled.

Given the many real threats that exist in the world, I suggest that any danger that is undetectable and unmeasurable should not play a role in policy. Adopt the Denver dose standard.

Beware

BEWARE OF FADS

Not all technologies will respond to optimism; you have to be innovative but realistic. For good scientific and engineering reasons, don't expect great advances in the hydrogen economy, geothermal, wave power, or tidal power. Biofuels have potential, but they have downsides too, particularly in the way they affect food supplies. Not all materials currently called biofuels truly help reduce greenhouse gases; after all, coal and oil were also created from plants and animals. Even if the right enzymes and microorganisms are developed, biofuels require large conversions of land use; their value will be measured over decades, not years.

BEWARE OF RISK-BENEFIT CALCULATIONS

Simplistic calculations of risks and benefits can be misleading. Did the United States' virtual moratorium on new nuclear power plants over the last 30 years result in saved lives? Or result in increased coal use that causes harm from the emission of mercury and sulfuric acid and other pollutants? How do we calculate the dangers of nuclear waste storage, and compare them to the dangers of, say, coal ash storage?

In the 1980s I resigned from the Sierra Club. In my letter (which I no longer have) I complained that the organization's opposition to nuclear power would drive the United States to using more fossil fuels, which could have dangerous implications for global warming. I was already familiar with that possibility, thanks to Roger Revelle and Gordon MacDonald. In 1980, MacDonald was the main author of the first detailed analysis of global warming; in 2000 he was my coauthor on a technical book on paleoclimate.

What is the indirect cost of carbon dioxide emissions? There is a consensus that carbon dioxide raises the temperature, but by how much?—and how much will that hurt? Or will it be good? Will the extra carbon dioxide have benefits, such as enhanced plant growth? Is there really any hope that we can calculate these effects? I honestly don't know, but it is important to keep these issues in mind. There is a real danger of tilting your logic in a predetermined direction to favor the answer you want. We do need to recognize that nothing is all bad or all good.

Beware of the Precautionary Principle

The *precautionary principle* sounds self-evident: always try to err on the side of safety. But what is safety? How do you balance environmental concerns with national security and economic well-being? You might be tempted to overemphasize those issues that are most evident to the public. Doing so may be politically precautionary, but it doesn't fulfill your leadership responsibilities. The problem is that the meaning of *precautionary* is in the mind of the advocate, and that makes the precautionary principle virtually useless in practice.

Beware of Optimism Bias and Skepticism Bias

Optimism bias is the belief that any technology can advance quickly; all it takes is sufficient effort and sufficient dollars. Frequently cited examples include the Manhattan Project and the rapid development of home computers. Beware. When you are president you can't afford to forget the counterexamples, the great programs that promised revolutionary results but produced only incremental ones: the war on cancer, the war on poverty, the war on drugs, the war on terror . . .

Skepticism bias is the trap of believing that for a technology you don't like, the problems will be insuperable: no, we can't prevent pollution from fracking; no, we'll never make sequestration work; no, nuclear will never be safe. It is often accompanied by optimism bias for the technology you like: yes, we can make solar cells cheaply; yes, we can develop batteries that can be recharged 10,000 times;

yes, we can find cost-effective ways to extract energy from waves. Optimism bias is evident nearly every time you see a headline about a technological "breakthrough" by a small company. Skepticism bias is evident whenever you hear sarcastic ridicule when a company such as BP says it's working hard to protect the environment.

Both skepticism bias and optimism bias will be hidden under conviction. People who feel strongly that their opinion is correct will emphasize certain facts and dismiss others. They're fooling themselves; don't let them fool you too. Claims based on conviction are not as valid as those based on objective analysis.

BEWARE OF APHORISMS

When I was a child, I was taught that *the solution to pollution is dilution*. I'm not kidding; I remember that little rhyme vividly. Now it sounds very naïve. People didn't seem to realize that the oceans (where we dumped our New York City garbage) or the atmosphere (where we dumped our sulfur and nitrous oxides and carbon dioxide) were finite and would soon show the effects.

Today there is an equally misleading mantra: "Think globally; act locally." I wish it would work. The problem is that local solutions don't always translate into global ones. Local solutions are often nothing more than feel-good measures. We can readily adopt expensive measures to reduce local pollution, but if they are irrelevant to the larger world, they don't address the real problem. An example is the all-electric lithium-ion auto.

Avoid referring to energy sources using feel-good words such as *green* and *renewable* and *clean*—all of which can be interpreted to rule out important technologies such as nuclear power, natural gas, and synfuels. Better to use *sustainable*, and to interpret that as "sustainable over the next 20–40 years," since we can't even guess at technology beyond those horizons. Even better is to use *alternative*, since our balance of payments and transportation energy security do make for a current crisis. Energy sources that are not sustainable over the very long term, such as shale gas, may be exactly what we need to bridge through a tough period to a great future.

Although there are many aphorisms that sound wise but don't really work, here's one that may be true: "To be truly sustainable, it must be profitable." Of course, profit could and should contain indirect costs, such as those to the environment. Unfortunately, there is no true consensus on how to measure those costs, and they are easily manipulated by advocates trying to make a case for their favorite approach. It is important to examine estimates of indirect costs carefully and objectively.

Here's another aphorism that I like—one that appeared earlier in this book: "A gallon saved is a gallon not imported."

Your Legacy

As soon as your term is over, you will be pronounced by some pundits to be the worst president in US history. At the same time, others will be comparing you to Washington and Lincoln. History, however, will eventually judge you not by the short-term impression you leave, but by your lasting accomplishments.

Your greatest energy challenge will be striking a balance between global warming and energy security. There will be pressure to take feel-good actions, to address short-term issues at the expense of long-term concerns. To earn your place in history, you have to have vision, trust in science and objective analysis, and think long-term.

You will have the world's toughest job. The buck will stop with you. Energy is contentious and misunderstood and politicized. You'll not only need to make the right decisions, but you'll need to educate the public so that they will recognize that you're making the right decisions. Your understanding of energy will be a great asset. Best of luck.

Notes

Preface

1. **[p. xv]** I've verified this correct attribution with both the Mark Twain Project at the University of California, Berkeley, and a descendant of Josh Billings.

Part I: Energy Catastrophes

Chapter 1: Fukushima

1. **[p. 19]** If 4,400 cancers are expected, ordinary statistical fluctuations are plus or minus 66 (the square root of 4,400) for one standard deviation; the expected increase is only 3 times this. Moreover, because of lifestyle differences, detecting an increase in the cancer rate typically requires subdividing the population, and when that's done the statistical significance drops dramatically.

2. **[p. 20]** This approximate conversion factor can be deduced from the summary paper "Health Risk Attributable to Environmental Exposures: Radon," by P. S. Steifer and B. R. Weir, published in the *Journal of Hazardous Material* volume 39 (1994), pages 211–223. The conversion is approximate because it depends on many factors, including the rate at which you breathe and the time before the radon decay products get to your body. The excess dose in Denver County averages 4.5 picocuries per liter; multi-

ply by 0.09 to get 0.4 rem per year. That's 0.3 rem per year higher than the US average radon exposure.

3. [p. 20] If you do a controlled study of a population of a million people (well beyond current medical capabilities), you expect about 200,000 cancers. But that number is unlikely to be exact; natural variations occur in any statistical process, typically about the square root of the number. So we expect the number 200,000 to be accurate to only the square root of 200,000, which is plus or minus 447. These statistical variations are larger than the number of excess cancers expected from the 0.3-rem exposure, which is .00012 × 1,000,000 = 120, so the cancer increase is statistically unobservable. In a study of 100 million people, the expected standard deviation is 4,470, and the number of cancers expected from the radiation is 12,000—so the cancer increase is barely detectable at a significance less than 3 standard deviations.

Chapter 2: The Gulf Oil Spill

4. [p. 26] "Sound Politics: Sound Commentary on Current Events in Seattle, Puget Sound and Washington State," July 15, 2010, http://soundpolitics .com/archives/014103.html.

5. [p. 30] The official number is 53,000 barrels per day. Since there are 42 gallons per barrel, and 86,400 seconds per day, this is equivalent to 26 gallons per second.

6. [p. 36] For a detailed breakdown, see the report "The Economic Cost of a Moratorium on Offshore Oil and Gas Exploration to the Gulf Region," by Joseph R. Mason, Louisiana State University, July 2010, available at http:// www.noia.org.

7. [p. 36] Kathy Finn, "Gulf Gets Taste of Recovery One Year after Spill," Reuters, April 20, 2011, http://www.reuters.com.

Chapter 3: Global Warming and Climate Change

8. [p. 40] We check these methods with modern data by ignoring regions that had no coverage back then, estimating the global temperature using the partial data, and then seeing how well the answer agrees with the correct answer that was derived by using all the data.

9. [p. 40] The UK group HadCRU is a collaboration between the Hadley Centre of the UK Met Office and the Climate Research Unit of the University of East Anglia. Part of this group became infamous in the "Climategate" scandal. The NASA group is located at the Goddard Institute for Space Science (GISS) in New York and is led by Jim Hansen, famous

for his outspoken warnings about the threat of climate change. NOAA stands for "National Oceanographic and Atmospheric Administration"; its climate team is located in Ashville, North Carolina. Our team, Berkeley Earth, involves scientists and statisticians at Berkeley, in Oregon, and in Georgia, assembled under the auspices of Novim, a nonprofit organization in Santa Barbara, California. About the same time that we were releasing our first results, Berkeley Earth team member Saul Perlmutter was awarded the 2011 Nobel Prize in Physics for his work in the discovery of dark energy.

10. **[p. 46]** At a climate conference in Santa Fe, New Mexico, in November 2011, I made a presentation that threw doubt on the volcanic interpretation of the dips. I pointed out that the cooling due to the 1815 Tambora eruption seemed to have begun in 1809. Maybe the "year without a summer" was due to something else, not Tambora. But that evening, Rohde went online and found a neglected publication showing that there had actually been another huge volcanic eruption 6 years prior to Tambora, exactly at our cold spike. This eruption was not known historically (and it is still unnamed), but the sulfate deposits in the ice cores for this explosion were comparable to those of Tambora. Instead of contradicting the belief that the cooling was caused by volcanic eruptions, the evidence now indicated that we could actually detect such eruptions just from the temperature record alone. On the next day at the conference, I was given five minutes to announce that my previous conclusion was wrong, and that the data actually confirmed the volcanic origin of the temperature dips.

11. **[p. 49]** The CO_2 levels for the preindustrial world, for the present world, and at double the preindustrial level, are 280, 395, and 560 parts per billion (ppb), respectively. When taking the logarithm, the *ppb* part can be ignored, since it will cancel when we take differences. Then the logarithms (to base 10; any base can be used) for these numbers are: 2.45, 2.60, and 2.75. Note that the logarithms increase by 0.15 for each step; this agreement is accidental but useful; it means that the expected temperature rise for each step is the same. We had a 1.6°C land temperature rise for the first step (1753 to 2012), so we expect a 1.6°C land temperature rise for the second step—that is, taking us from 395 ppb to 560 ppb, which will occur in 2052, if the exponential increase in CO_2 continues.

12. **[p. 49]** Mathematically, we say that $\log(\exp(x)) = x$, where *log* stands for the natural logarithm. The equivalent statement using base 10 logarithms is $\log_{10}(10^x) = x$.

13. **[p. 56]** Chris Landsea's chart can be found in the article titled "Counting

Atlantic Tropical Cyclones back to 1900," published in the journal *Eos*, volume 88 (May 2007), pages 197–202.

14. **[p. 65]** Much of this section was adopted from my *Wall Street Journal* Op-Ed article "Naked Copenhagen," published December 12, 2009, in the midst of the Copenhagen meeting.

15. **[p. 66]** J. G. J. Olivier, G. Janssens-Maenhout, J. A. H. W. Peters, and J. Wilson, *Long-Term Trends in Global CO$_2$ Emissions*. *2011 Report* (The Hague, Netherlands: PBL Netherlands Environmental Assessment Agency, 2011).

16. **[p. 67]** The data for this figure come from *Long-Term Trends in Global CO$_2$ Emissions*. *2011 Report*, by J. G. J. Olivier, G. Janssens-Maenhout, J. A. H. W. Peters, and J. Wilson (The Hague, Netherlands: PBL Netherlands Environmental Assessment Agency, 2011).

17. **[p. 74]** The addition of the term *auditor* was suggested by Judith Curry. Steve McIntyre, one of the most careful of the skeptics, describes himself in that way.

18. **[p. 75]** Recall that the consensus IPCC conclusion is that *most* of the warming of the past 50 years was due to humans. Thus, it is technically possible to agree fully with the IPCC and yet disagree that *all* global warming (starting back in the 1850s) was caused by humans.

Part II: The Energy Landscape

1. **[p. 83]** In 2008 the US population was 305 million.

2. **[p. 83]** See *A Cubic Mile of Oil: Realities and Options for Averting the Looming Global Energy Crisis*, by H. D. Crane, E. M. Kinderman, and R. Malhotra (Oxford, England: Oxford University Press, 2010).

CHAPTER 4: THE NATURAL-GAS WINDFALL

3. **[p. 89]** "Saudi Prince Seeks to Discourage Western 'Alternatives' to High Priced Oil," *International Business Times*, May 30, 2011.

4. **[p. 97]** US Energy Information Administration, *World Shale Gas Resources: An Initial Assessment of 14 Regions outside the United States* (Washington, DC: US Department of Energy, 2011).

5. **[p. 99]** Water that is chilled on the surface contracts and sinks to the bottom. So the coldest water is found on the seafloor. However, water at the bottom is not cold enough to freeze, because once the temperature drops

below 4°C (39°F) the water expands slightly, becomes less dense, and floats upward. So at the seabed the water is typically 4°C.

Chapter 5: Liquid Energy Security

6. **[p. 105]** A gallon of oil delivers 33.7 kilowatt-hours of energy. If you pump 10 gallons in 5 minutes, you are providing 337 kilowatt-hours every one-twelfth of an hour; that's 337 × 12 = 4,040 kilowatts = 4 megawatts. Typical homes in the United States use between 1 and 1.5 kilowatts of electricity, on average.

7. **[p. 106]** A somewhat more sophisticated approach to natural-gas supply is called the "resource triangle." For an introduction, see the National Petroleum Council's Topic Paper no. 29, titled *Unconventional Gas* (Working Document of the NPC Global Oil & Gas Study, July 18, 2007), www.npc .org/study_topic_papers/29-ttg-unconventional-gas.pdf.

8. **[p. 106]** In 1975, President Ford abolished the old Atomic Energy Commission and replaced it with the Energy Research and Development Administration (ERDA). Then, in 1977, the organization reached Cabinet status when President Carter converted it to the Department of Energy.

Chapter 7: Energy Productivity

9. **[p. 120]** Since there are 8,766 hours in a year, the plant would sell 8,766 gigawatt-hours. A cost of 15¢ per kilowatt-hour is the same as $150,000 per gigawatt-hour, so the revenue would be 8,766 × $150,000 = $1.3 billion.

10. **[p. 123]** Heat radiation is called "infrared" by scientists. Visible light also carries heat, but it is not usually included in the term *heat radiation*.

11. **[p. 129]** Paul Hawken, Amory Lovins, and L. Hunter Lovins, *Natural Capitalism: Creating the Next Industrial Revolution* (Boston: Little, Brown, 1999), p. 245.

12. **[p. 131]** Don't confuse these legitimate photos with the widely circulated phony ones. For the distinction, see www.snopes.com/photos/space/ blackout.asp.

13. **[p. 136]** MeV stands for "million electron-volts." One MeV is the energy that an electron gets when it jumps across wires that have a million volts of electric potential. When that electron gives all that energy to a gamma ray, the gamma ray has 1 MeV of energy. One MeV is a typical energy for particles and rays coming out of a nucleus. The energy from a typical particle of light coming from an atom is typically a million times smaller: 1 eV (1 electron-volt).

Part III: Alternative Energy

CHAPTER 8: SOLAR SURGE

1. **[p. 145]** Actually, for the same power, sunlight is about seven times brighter than ten 100-watt bulbs. That's because the bulbs emit more of their light at the red/infrared end of the spectrum, and the eye is insensitive to these wavelengths. A fair comparison of the brightness of tungsten filament lightbulbs to that of sun requires the concept of *luminous efficacy*—a measure of brightness perceived by the human eye. The sun has a luminous efficacy of 13.6%, and a tungsten bulb has 2%. That means that the light from the tungsten bulb, for the same wattage, is about 2/13.6 = 15% as bright. A compact fluorescent bulb has a luminous efficacy of about 10%, five times greater than that of a tungsten bulb, but still lower than that of the sun. A green laser has nearly a 100% luminous efficacy.

2. **[p. 146]** There's a quick way to see the factor of one-quarter, if you know some geometry. The cross-sectional area of the Earth—that is, the projected area toward the sun—is πR^2, where R is the Earth's radius. That tells you how much sunlight the Earth absorbs. But the total surface area of a sphere is $4\pi R^2$, and as the Earth turns, the energy is spread out over this area, which is 4 times larger. So, on average, each part of the Earth receives only ¼ of what it gets at peak. On the equator, that factor is ¹⁄π of the peak (the ratio of the diameter of the equator to its circumference).

3. **[p. 147]** Please don't write me to say that it is really the Earth that is moving, not the sun. In fact, in physics we often use accelerated or rotating coordinate systems (although we avoid these in elementary physics), and in such a system it actually is the sun that is moving. For this reason, in advanced physics, Tycho Brahe was just as right as Nicolaus Copernicus.

4. **[p. 150]** The efficiency of conversion is limited by a basic law of physics, known as the Carnot efficiency, which is given by this equation: Efficiency $= 1 - T_{cold}/T_{hot}$, where T_{hot} is the temperature of the salt, measured on the absolute temperature scale (either Fahrenheit or Celsius is OK). To convert Fahrenheit to absolute, add 459 degrees; to convert Celsius to absolute, add 273 degrees. So, for example, if the salt is heated to 1,000°F by the solar energy, then T_{hot} = 1,459. T_{cold} is typically the temperature of the environment, perhaps 70°F + 459 = 529. That means the efficiency is

$$\text{Efficiency} = 1 - \frac{529}{1459} = 0.637 = 63.7\%$$

That's impressive. The very best (super expensive) solar cells reach only 42% efficiency, and the cheaper variety can achieve only 15%. Wow. Since some heat is lost to the environment, your efficiency will be lower.

5. [p. 151] As an example, the 1-megawatt plant built by Suntech in China proudly delivers about 1,000 megawatt-hours of energy per year. But there are 8,766 hours per year, so if it constantly ran at its 1-megawatt peak, it would give 8,766 megawatt-hours—8.7 times as much energy as it actually produces.

6. [p. 152] It is called an inverter because of the way it works: every sixtieth of a second it inverts positive voltage to negative voltage (for $\frac{1}{120}$ of a second), thus converting the *direct* current (DC) to 60-cycle-per-second *alternating* current (AC). Then the voltage can be raised or lowered by transformers, which work only with AC. A "grid tie" inverter is one that sets its frequency to match that of the electric grid, so that power can be injected into the grid.

CHAPTER 9: WIND

7. [p. 159] The energy carried by the air increases as the square of the velocity. (That's just the kinetic energy $E = \frac{1}{2}mv^2$ law.) But the rate at which this energy reaches you is also proportional to v. So the power is proportional to $v^2 \times v = v^3$.

8. [p. 162] If you know a little bit of physics and are interested in learning more, you can find a clear derivation of this law in the Wikipedia article "Betz' law," online.

CHAPTER 10: ENERGY STORAGE

9. [p. 170] Lead oxide and lead sulfate are also conductors, and they play critical roles in the charge and discharge cycles.

10. [p. 170] C. Daniel and J. O. Besenhard (Eds.), *Handbook of Battery Materials* (2nd ed.; 2 vols.) (Weinheim, Germany: Wiley-VCH, 2011).

11. [p. 172] The simple relationship between gas storage and tank weight is a nice exercise for an undergraduate physics or engineering student. For simplicity, consider a hollow spherical tank of radius R and thickness T holding gas at pressure P. Imagine that the two halves of the tank are held together by metal with area $2\pi RT$. The force of separation is given by the area multiplied by the pressure: $F = \pi R^2 P$. If the metal is at its maximum strength (with some safety factor), the force holding the two sides together is proportional to the area of the metal holding it together: $F \sim 2\pi RT$. The force of separation is $\pi R^2 P$, so setting these equal, we find that T is propor-

tional to PR: $T \sim PR$. The mass of gas in the tank (M_g) is proportional to the volume multiplied by the pressure: $M_g \sim \frac{4}{3}\pi R^3 P$. The mass of the tank ($M_t$) is proportional to the thickness T multiplied by the surface area of the sphere: $M_t \sim T4\pi R^2$. The ratio of the mass in the tank to that in the container is therefore proportional to $(R^3 P)/(TR^2)$. Substitute $T = PR$, and the size cancels.

12. **[p. 172]** To do this calculation, the two laws you need are $PV = nkT$ and the adiabatic equation: PV^γ = constant, where $\gamma = 1.4$ for air.

13. **[p. 172]** The expanding gas will refrigerate the room. This is fun to see with a carbon dioxide fire extinguisher. Release the gas and not only does the cylinder get cold, but the escaping gas will make snow from water vapor in the air. Be sure to recharge the tank after you've tried this.

14. **[p. 175]** Hoop stress is proportional to the centrifugal force, which is proportional to the mass multiplied by the velocity squared, divided by the radius. But the mass is proportional to the circumference of the hoop, also proportional to the radius. So the radius cancels out. Therefore, the maximum speed of the hoop material does not depend on the radius, and so neither does the energy stored per unit mass.

Chapter 11: The Coming Explosion of Nuclear Power

15. **[p. 181]** If you know high school chemistry, then you can follow this calculation. A 1.4-pound sample of uranium is 636 grams. One mole of U-235 weighs 235 grams, so our 1.4 pounds contains 2.7 moles. Every mole contains an Avogadro's number worth of atoms, 6×10^{23}, so altogether our uranium has 1.6×10^{24} atoms. With a pocket calculator (or a spreadsheet) you can verify that 2^{80} is approximately equal to this number. So, 80 stages will explode all the atoms.

16. **[p. 181]** The term *shake* is attributed to Enrico Fermi. According to legend, he got it from the old phrase "quick as two shakes of a lamb's tail." Fermi also named the unit for the surface area of a nucleus: the *barn*, equal to 10^{-24} square centimeter. Aiming neutrons at atoms was "as easy as hitting the side of a barn."

17. **[p. 181]** The number in U-238 comes from the total number of protons and neutrons in the nucleus: $92 + 146 = 238$. For U-235 the numbers are $92 + 143 = 235$.

18. **[p. 183]** The cooling water at Chernobyl also acted as a moderator, although it was not the primary one; yet its tendency to form gaseous voids contributed to the instability of the reactor.

19. **[p. 183]** To differentiate ordinary water from heavy water, it's called "light water." So the United States uses "light water reactors." I guess they didn't want to call them "ordinary water reactors."

20. **[p. 184]** Most of the downtime is for scheduled maintenance; if not for this, the capacity factor would be even higher than 90%.

21. **[p. 185]** Two firefighters died on the night of the accident, and 28 more over the next few weeks from radiation illness.

22. **[p. 188]** Plutonium-239 is also a fissionable target for most reactors. Even if it is not specifically introduced as a fuel, it is created in the reactor when a neutron is absorbed by U-238. This plutonium then serves as an additional fuel for the chain reaction.

23. **[p. 188]** There are 43 zeros after the decimal point in this number. Numbers like this illustrate why scientists love scientific notation.

24. **[p. 189]** Here's a technical explanation for physics majors using some nuclear physics jargon: The increased absorption by U-238 is a result of a strong absorption resonance near the energy of 1 electron-volt. Higher temperature results in Doppler broadening, and it is the tail of this resonance that causes the absorption.

25. **[p. 190]** A more sophisticated calculation takes into account the fact that, as the uranium gets enriched it gets harder to enrich further, since there are fewer and fewer U-238 atoms to remove. If you're interested in the details, look up *SWU*, an acronym for "separative work unit." But the simplified calculation I do in the text illustrates why getting to 19.9% enrichment is much harder than taking it from there to 90%.

26. **[p. 191]** Many people think that Saddam Hussein never enriched uranium, but they are confusing the second Gulf War with the first Gulf War. After the first Gulf War, the UN determined that Saddam had created a multibillion-dollar effort to enrich uranium. He had a calutron facility similar in capability to that of the US World War II Manhattan Project, and he may have already used it to enrich small amounts of uranium to weapons grade. The calutrons he built were destroyed by the UN. For the UN report see www.fas.org/news/un/iraq/iaea/s1998694.htm. For photos of the destroyed calutrons, see: www.fas.org/nuke/guide/iraq/nuke/program.htm.

27. **[p. 192]** The original report is 1,016 pages. For a popular account of Deffeyes and MacGregor's work, see their article "World Uranium Resources," *Scientific American* 242(1) (January 1980): 66–76.

Chapter 12: Fusion

28. [p. 199] Bhabha made this prediction in his presidential address at a United Nations "Atoms for Peace" conference in Geneva in 1955. For a transcript, see *Bhabha and His Magnificent Obsessions*, by G. Venkataraman (Hyderabad, India: Universities Press, 1994), p. 152.

29. [p. 200] The first-generation fusion plants will probably use deuterium and tritium as their fuel, not ordinary hydrogen. Deuterium is "heavy hydrogen," made heavier by an extra neutron in the nucleus. Even though deuterium is only 1/6,240 of ordinary hydrogen, it can be cheaply separated from water. Tritium is extra-heavy hydrogen (two extra neutrons), and it is very rare; because it is radioactive with a half-life of only 12 years, there are only about 16 pounds total in the oceans. But tritium can be manufactured in a fusion plant from the neutrons that are emitted; that requires lithium as a target.

30. [p. 202] The surface of the sun, which gives us all the light, is much much cooler, only about 6 *thousand* degrees Celsius. But it gets all its energy from the inner, hot part of the sun. No fusion takes place on the surface; it is too "cool."

31. [p. 207] In 1972, scientists at the Lawrence Livermore National Laboratory published a seminal paper showing that laser-induced implosion could lower the required laser energy needed to ignite fusion. The authors were John Nuckolls, Lowell Wood, Albert Thiessen, and George Zimmerman. The article was "Laser Compression of Matter to Super-High Densities: Thermonuclear (CTR) Applications," *Nature* 239 (1972): 139–142.

32. [p. 207] Here are the numbers. If a NIF-like laser is to produce 1 gigawatt of electric power, it must produce 3 gigawatts of heat. (Conversion is about 33% efficient.) A gigawatt is a million kilowatts, so in one hour, it will deliver 3,000,000 kilowatt-hours of electric energy. That's *per hour*; in $\frac{1}{10}$ second (the time for each minibomb), it delivers 1/36,000 of that—about 83 kilowatt-hours. Electricity currently costs the consumer an average of 10¢ per kilowatt-hour; that means each little hydrogen bomb delivers $8.30 worth of electricity. (Note that we're talking about a future laser system. The current NIF target is designed to deliver only 4 megawatts per pulse; that's about 100 kilowatt-hours, worth $1 retail.) So to be commercial, the laser approach has to cost less than $8 per shot, including interest on the investment. The engineers are convinced that this is possible. Right now the current target costs thousands of dollars to manufacture, but they think a simplified target can be made for less than

50¢. They consider $1 to be the threshold they have to reach. I have my doubts about whether they can make targets for so little, but if indeed a very simple target based on direct laser drive (no intermediate X-rays) can be built, then this approach may be a viable power source in a few decades.

33. **[p. 209]** Some neutrons are produced by hydrogen-boron collisions, but only about 0.1% of the time.

34. **[p. 210]** In elementary quantum physics, you can show that the size of the orbit is inversely proportional to the mass of the negative particle. That means that a muon gets 207 times closer to the nucleus than does an electron.

35. **[p. 211]** Sticking to the helium is more limiting than the fact that the muon lifetime is 2 microseconds.

36. **[p. 211]** On the nuclear scale, physicists like to use a simple unit of energy called the MeV, which stands for "million electron-volts." Don't worry about this unit, but let me use it to illustrate muon creation. To make a muon, you first make a particle called a pion; that takes a minimum of 140 MeV of energy (that's the rest mass energy from $E = mc^2$). Typically, one-third of these pions—the ones with negative charge—turn into the right kind of muon. So you need $3 \times 140 = 420$ MeV. Every time a muon catalyzes fusion, it releases, on average, 3.65 MeV of energy. So you reach the "scientific breakeven" point if the muon catalyzes $420/3.65 = 115$ fusions. But unfortunately that is heat energy, and to turn it into the kind of electric energy needed to create pions is inefficient. (The usual way of making pions is to accelerate a proton to at least 140 MeV of energy, and then collide it into a nucleus.) Because of the inefficiency, the minimum energy needed is about 3 times larger. That means you need to make 3 times as many fusions for each muon—about 350.

37. **[p. 214]** Heavy water is H_2O with the hydrogen replaced by heavy hydrogen—that is, hydrogen with an extra neutron in the nucleus. Such hydrogen is called deuterium, and we give it the symbol D. Electrolysis is the splitting of $2H_2O$ into $2H_2$ and O_2. For heavy water, the analogous reaction is $2D_2O$ splitting into $2D_2$ and O_2. The electrodes (the wires that deliver the electricity into the liquid) were made of palladium. Palladium is often used to catalyze chemical reactions, and Pons and Fleishman were hoping it could catalyze fusion.

38. **[p. 214]** I continued the chain reaction by faxing my copy of Pons and Fleischmann's paper to others. But first, on the front page, I affixed this sentence: "If you receive this paper, please send $1 to Richard Muller." I figured the paper would be copied and faxed exponentially. Several people

later told me that they saw that sentence on the paper. But nobody ever sent me $1. Shareware doesn't reap much reward.

39. **[p. 217]** You can watch this program at www.cbsnews.com/video/watch/ ?id=4955212n?source=mostpop_video.

Chapter 13: Biofuels

40. **[p. 221]** Timothy Searchinger, "A Quick Fix to the Food Crisis: Curbing Biofuels Should Halt Price Rises," *Scientific American*, July 2011, http://www.scientificamerican.com/article.cfm?id=a-quick-fix-to-the-food-crisis.

41. **[p. 226]** The numbers appear elsewhere in this book, but let me repeat them here. According to the IPCC, humans have caused an excess temperature rise of "most" of 0.64°C. Let's assume that *most* means 80%; in that case, the human-caused rise is 0.5°C. The United States has contributed about 20% of that, or 0.1°C. Our autos have contributed roughly one-quarter of that 0.1°C, about 0.025°C, or (expressed differently) ¼₀°C. With just modest effort, including improved CAFE (Corporate Average Fuel Efficiency) standards, we could easily keep the expected rise over the next 50 years to another ¼₀°C.

Chapter 14: Synfuel and High-Tech Fossil Fuels

42. **[p. 228]** Thomas W. Findley, my father-in-law, proposes that we use acronyms these days because we no longer study Greek and Latin and therefore don't know how to construct new words from classical language roots. When he named his newly invented glue *epoxy*, he created the word from the Greek root *ep* ("strong") and *oxy* from "oxygen."

43. **[p. 234]** That's simply a consequence of the fact that each atom of carbon makes one molecule of carbon dioxide, but the solid is much denser than the gas.

Chapter 15: Alternative Alternatives

44. **[p. 237]** Hydrogen can gain another factor of 1.7 in fuel efficiency if it is used in a fuel cell to drive an electric motor, rather than in an internal combustion engine. The tank-to-wheel efficiency of fuel cell vehicles has been shown to have an average value of 36% when tested under the "New European Driving Cycle." The comparable value for a diesel auto is 22%. Honda claims that its (very expensive) Clarity can achieve better tank-to-wheel efficiency—60%—but we need to be wary; recall that Toyota once claimed 60 miles per gallon for its Prius, a value that no driver ever actually reached, to my knowledge.

45. **[p. 243]** We can calculate the power from the mass m of the water (in kg) and the height h of the rise (in meters), using the physics formula that the energy is mgh, and $g = 9.8$ is the gravitational constant. Power is this energy (in joules) divided by the number of seconds in the half day. Put in the mass in kilograms (height × area × 1,000 kg/m³), and divide by 43,200 seconds in 12 hours, to get that the power available from tides in the basin is about 300 megawatts. The power plant manages to convert 100 megawatts of this into electricity.

46. **[p. 244]** The energy available for each cycle depends on the height of the water and on the amount of water—which is also proportional to the height. So the energy available, and the average power, depends on the height squared.

CHAPTER 16: ELECTRIC AUTOMOBILES

47. **[p. 248]** A gallon of gasoline contains an energy of about 33 kilowatt-hours. At a fill rate of 2 gallons per minute, that's half a minute, or 1/120 hour, per gallon, and the transfer is therefore (33,000 watt-hours) ÷ (1/120 hour) = 3.96×10^6 watts = 3.96 megawatts.

48. **[p. 249]** Over the past 50 years, the IPCC estimates that human-caused global warming has accounted for "most" of the 0.64°C rise that has been observed. A plausible estimate for the word *most* is 80%, so let's assume the human component was 0.5°C. Of this, the United States has contributed considerably more than any other country—about 20%. That makes the US contribution to global warming about 0.1°C. Our automobiles and trucks and other vehicles contributed about a quarter of this, or about 0.025°C = ¼₀°C.

49. **[p. 258]** The specs on the battery say it provides 273 volts and 6.5 amp-hours. We multiply volts by amp-hours to get watt-hours, so this battery provides 273 × 6.5 = 1,774 watt-hours = 1.774 kilowatt-hours.

50. **[p. 259]** I assumed $40 per pound and 40 watt-hours per pound; that's $1 per watt-hour, and $1,000 per kilowatt-hour.

51. **[p. 259]** The Corporate Average Fuel Economy, CAFE, is a limit on the average fuel economy of any manufacturer's fleet of cars and light trucks for a given year.

CHAPTER 17: NATURAL-GAS AUTOMOBILES

52. **[p. 261]** The current consumer price for natural gas (from my home bill) is $1.20 per therm. A therm is 29.3 kilowatt-hours, so the cost is 4¢ per kilowatt-hour. A gallon of gasoline, costing $3.50, can deliver 33.7

kilowatt-hours. So the energy from gasoline costs $3.50/33.7 = 10.4¢ per kilowatt-hour, about 2.5 times more than energy from natural gas. The discrepancy at wholesale prices is greater.

53. **[p. 264]** The density of natural gas at 1 atmosphere is 0.8 kilogram per cubic meter; at 250 atmospheres it is 200 kilograms per cubic meter. That is one-quarter the density of gasoline. However, natural gas delivers 30% more energy per kilogram, so the difference is about one-third.

CHAPTER 18: FUEL CELLS

54. **[p. 268]** A heat pump uses a little bit of energy to "pump" heat from the cold outdoor air into the building. A refrigerator does this; it pumps heat from within the refrigerator into the room. You can pump a lot of heat with just a little bit of energy, so if you measure heat provided and divide that number by energy used, you get an efficiency greater than 1. If efficiency greater than 1 sounds mysterious, consider this analogy: it takes very little of your physical energy to pump gasoline from a service station. The heat pump is doing a similar thing: using a small amount of energy to move a large amount of energy from one place to another.

55. **[p. 269]** The Y stands for "yttrium oxide" (yttrium is an exotic element with atomic number 39 on the periodic table). The Z is for an oxide of "zirconium," itself atomic number 40. The S is for "stabilized." The full acronym YSZ stands for "yttria-stabilized zirconia." Stabilization is necessary because the crystal structure of zirconia changes when it is heated, the size of the crystal changes, and the ceramic is likely to crack. Yttrium oxide stabilizes zirconia by suppressing the change in structure.

56. **[p. 270]** "'This Is Brand New': The CEO of Bloom Energy on a New Way of Powering the Planet" (in *Newsweek* magazine), *The Daily Beast*, April 22, 2010, http://www.thedailybeast.com/newsweek/2010/04/22/this-is-brand-new.html.

57. **[p. 270]** US Energy Information Administration, "Updated Capital Cost Estimates for Electricity Generation Plants," November 2010, http://205.254.135.24/oiaf/beck_plantcosts.

CHAPTER 19: CLEAN COAL

58. **[p. 273]** That's because coal is mostly carbon, and when it burns, most of the energy comes from creating carbon dioxide (CO_2). When methane (CH_4) burns, half the energy comes from burning the hydrogen to make water—so that half of the energy production creates no CO_2.

59. **[p. 275]** To work the calculation, you need to know (from Table IV.1),

that one pound of coal produces 3.2 kilowatt-hours of heat energy, that a reasonable efficiency for a power plant (heat energy to electricity) is 31%, that there are 2,000 pounds per ton, and 3,600 seconds in an hour. Here's the calculation: One ton of coal (2,000 pounds) can produce $3.2 \times 2,000$ = 6,400 kilowatt-hours of heat = 6.4 megawatt-hours of heat = $6.4 \times 31\%$ = 2 megawatt-hours of electricity. That means it can deliver 2 megawatts for 1 hour (3,600 seconds), or 7.2 gigawatts for 1 second. Equivalently, one ton of coal can deliver 1 gigawatt for 7.2 seconds. Assume for simplicity that coal is pure carbon. When a ton of carbon combines with oxygen to form CO_2, the combination weighs 3.7 tons. It produces 1 ton of CO_2 in $7.2/3.7 = 1.95$ seconds.

60. **[p. 275]** Old king coal is not a merry old soul . . . (sorry).

Part IV: What Is Energy?

1. **[p. 282]** The table does not include the weight of the oxygen that combines with some of the fuels. If you prefer kilograms in place of pounds, just double the numbers (1 kilogram = 2.2 pounds). TNT energy is defined for arms control purposes to be 1 kilocalorie per gram, but I am using a number that reflects its value in actual explosive use.

2. **[p. 285]** If the mass of the object at rest is m_0, then the "kinetic mass" m is m_0 divided by $\sqrt{1 - (v/c)^2}$, where v is the velocity of the object and c is the speed of light. But be careful: Newton's law $F = ma$ is no longer true, even if you use this new mass.

3. **[p. 287]** When I say that the true equations of physics don't change with time, I am not referring to our knowledge of those equations. In 1900, we didn't know that $E = mc^2$. A few years later we did. The physics equations didn't change; only our knowledge of those equations did.

4. **[p. 289]** For classical motion (a moving mass), the new momentum turns out to be equal to the old momentum: $m_0 v$, divided by $\sqrt{1 - (v/c)^2}$, where v is the velocity of the object and c is the speed of light. This combination is sometimes called the *relativistic* momentum.

5. **[p. 290]** In the jargon of physics, that makes energy and time *conjugate variables*. Position and momentum are also conjugate variables.

CREDITS

I.1. Satellite view of the Fukushima Dai-ichi nuclear power plant. By DigitalGlobe via Getty Images. Editorial image # 110051644.

I.2. Photograph of Chernobyl. Courtesy of Soviet authorities via Wikipedia.

I.3. First-year dose estimate map. Courtesy of the National Nuclear Security Administration.

I.4. Photograph of oil-soaked pelican by Charlie Reidel for the Associated Press. Image ID# 100603064872.

I.5. Map of the *Deepwater Horizon* oil spill. Courtesy of the *New York Times*.

I.6. Graph of decadal land-surface average temperature. Courtesy of Berkeley Earth (www.BerkeleyEarth.org).

I.7. Map showing temperature stations with increases and decreases in temperature over the last century. Courtesy of Berkeley Earth (www.BerkeleyEarth.org).

I.8. Graph showing the number of hurricanes over the past 150 years. Courtesy of the author.

I.9. Plot comparing the Atlantic hurricane seasons of 1933 and 2005. Published by Chris Landsea, NOAA, National Hurricane Center, Miami, Florida.

I.10. Graph showing the number of strong-to-violent tornadoes in the United States in March–August, 1950–2010. Courtesy of NOAA.

I.11. Composite images of Arctic ice in 1979 and 2003, from satellite images analyzed by the NASA Goddard Scientific Visualization Studio.

I.12. Photograph of the *Gjøa*. Public domain via Wikipedia.

I.13. Graph showing the temperature of the Earth going back 1,000 years. Courtesy of the World Meteorological Association.

I.14. Original graph showing the temperature of the Earth going back 1,000 years. Courtesy of the World Meteorological Association.

I.15. Plot showing 23 annual tide gauge records. Courtesy of Robert Rohde, Global Warming Art.

I.16. Graph showing the growth in CO_2 emissions following the Copenhagen "success." Courtesy of the author and associates.

II.1. Plot comparing primary energy per capita and GDP per capita. Courtesy of the author.

II.2. Plot showing the US total energy flow, 2010. Courtesy of the US Energy Information Administration, http://www.eia.gov/totalenergy/data/annual/diagram1.cfm.

II.3. Plot showing the average US production of shale per year. Courtesy of the US Energy Information Administration.

II.4. Photograph of Saudi Prince Al-Waleed bin Talal by Fahad Shadeed for Reuters. Image ID# GM1E7391TRC01

II.5. Photograph of a natural-gas flare in Thailand. Creative Commons, http://en.wikipedia.org/wiki/File:PTT_flame_1.jpg.

II.6. Schematic of horizontal drilling into a shale gas formation. From Stephen Marshak, *Earth: Portrait of a Planet* (New York: W. W. Norton & Co., 2007).

II.7. Map showing shale gas plays in the lower 48 United States. Courtesy of the US Energy Information Administration.

II.8. Map showing 48 shale gas basins in 32 countries. From "World Shale Gas Resources: An Initial Assessment of 14 Regions Outside the United States" by the US Energy Information Administration, http://www.eia.gov/analysis/studies/worlsdhalegas/.

II.9. Photograph of methane hydrate. Courtesy J. Pinkston and L. Stern / US Geological Survey.

II.10. Map showing methane hydrate locations. Courtesy of the United States Geological Survey.

II.11. Graphs showing reserves of oil, natural gas, and coal. Courtesy of the author.

II.12. Photograph of Jimmy Carter wearing a sweater.

INDEX

Page numbers in *italics* refer to figures and tables.